T0298045

CAMBRIDGE STUDIES IN ADVANCED MATHEMATICS 128

ZETA FUNCTIONS OF GRAPHS

Graph theory meets number theory in this stimulating book. Ihara zeta functions of finite graphs are reciprocals of polynomials, sometimes in several variables. Analogies abound with number-theoretic functions such as Riemann or Dedekind zeta functions. For example, there is a Riemann hypothesis (which may be false) and a prime number theorem for graphs. Explicit constructions of graph coverings use Galois theory to generalize Cayley and Schreier graphs. Then non-isomorphic simple graphs with the same zeta function are produced, showing that you cannot "hear" the shape of a graph.

The spectra of matrices such as the adjacency and edge adjacency matrices of a graph are essential to the plot of this book, which makes connections with quantum chaos and random matrix theory and also with expander and Ramanujan graphs, of interest in computer science. Pitched at beginning graduate students, the book will also appeal to researchers. Many well-chosen illustrations and exercises, both theoretical and computer-based, are included throughout.

Audrey Terras is Professor Emerita of Mathematics at the University of California, San Diego.

CAMBRIDGE STUDIES IN ADVANCED MATHEMATICS

All the titles listed below can be obtained from good booksellers of from Cambridge University Press. For a complete series listing visit: http://www.cambridge.org/series/sSeries.asp?code=CSAM

Already published

78 V. Paulsen *Completely bounded maps and operator algebras*
79 F. Gesztesy & H. Holden *Soliton equations and their algebro-geometric solutions, I*
81 S. Mukai *An introduction to invariants and moduli*
82 G. Tourlakis *Lectures in logic and set theory, I*
83 G. Tourlakis *Lectures in logic and set theory, II*
84 R. A. Bailey *Association schemes*
85 J. Carlson, S. Müller-Stach & C. Peters *Period mappings and period domains*
86 J. J. Duistermaat & J. A. C. Kolk *Multidimensional real analysis, I*
87 J. J. Duistermaat & J. A. C. Kolk *Multidimensional real analysis, II*
89 M. C. Golumbic & A. N. Trenk *Tolerance graphs*
90 L. H. Harper *Global methods for combinatorial isoperimetric problems*
91 I. Moerdijk & J. Mrcun *Introduction to foliations and Lie groupoids*
92 J. Kollár, K. E. Smith & A. Corti *Rational and nearly rational varieties*
93 D. Applebaum *Lévy processes and stochastic calculus (1st edition)*
94 B. Conrad *Modular forms and the Ramanujan conjecture*
95 M. Schechter *An introduction to nonlinear analysis*
96 R. Carter *Lie algebras of finite and affine type*
97 H. L. Montgomery & R. C. Vaughan *Multiplicative number theory, I*
98 I. Chavel *Riemannian geometry (2nd edition)*
99 D. Goldfeld *Automorphic forms and L-functions for the group GL(n,R)*
100 M. B. Marcus & J. Rosen *Markov processes, Gaussian processes, and local times*
101 P. Gille & T. Szamuely *Central simple algebras and Galois cohomology*
102 J. Bertoin *Random fragmentation and coagulation processes*
103 E. Frenkel *Langlands correspondence for loop groups*
104 A. Ambrosetti & A. Malchiodi *Nonlinear analysis and semilinear elliptic problems*
105 T. Tao & V. H. Vu *Additive combinatorics*
106 E. B. Davies *Linear operators and their spectra*
107 K. Kodaira *Complex analysis*
108 T. Ceccherini-Silberstein, F. Scarabotti & F. Tolli *Harmonic analysis on finite groups*
109 H. Geiges *An introduction to contact topology*
110 J. Faraut *Analysis on Lie groups: an introduction*
111 E. Park *Complex topological K-theory*
112 D. W. Stroock *Partial differential equations for probabilists*
113 A. Kirillov, Jr *An introduction to Lie groups and Lie algebras*
114 F. Gesztesy *et al. Soliton equations and their algebro-geometric solutions, II*
115 E. de Faria & W. de Melo *Mathematical tools for one-dimensional dynamics*
116 D. Applebaum *Lévy processes and stochastic calculus (2nd edition)*
117 T. Szamuely *Galois groups and fundamental groups*
118 G. W. Anderson, A. Guionnet & O. Zeitouni *An introduction to random matrices*
119 C. Perez-Garcia & W. H. Schikhof *Locally convex spaces over non-Archimedean valued fields*
120 P. K. Friz & N. B. Victoir *Multidimensional stochastic processes as rough paths*
121 T. Ceccherini-Silberstein, F. Scarabotti & F. Tolli *Representation theory of the symmetric groups*
122 S. Kalikow & R. McCutcheon *An outline of ergodic theory*
123 G. F. Lawler & V. Limic *Random walk: a modern introduction*
124 K. Lux & H. Pahlings *Representations of groups*
125 K. S. Kedlaya p-*adic differential equations*
126 R. Beals & R. Wong *Special functions*
127 E. de Faria & W. de Melo *Mathematical aspects of quantum field theory*

Zeta Functions of Graphs

A Stroll through the Garden

AUDREY TERRAS
University of California, San Diego

CAMBRIDGE
UNIVERSITY PRESS

University Printing House, Cambridge CB2 8BS, United Kingdom

One Liberty Plaza, 20th Floor, New York, NY 10006, USA

477 Williamstown Road, Port Melbourne, VIC 3207, Australia

4843/24, 2nd Floor, Ansari Road, Daryaganj, Delhi - 110002, India

79 Anson Road, #06-04/06, Singapore 0799060

Cambridge University Press is part of the University of Cambridge.

It furthers the University's mission by disseminating knowledge in the pursuit of education, learning and research at the highest international levels of excellence.

www.cambridge.org
Information on this title: www.cambridge.org/9780521113670

First published 2011

A catalogue record for this publication is available from the British Library

Library of Congress Cataloging in Publication data
Terras, Audrey.
Zeta functions of graphs : a stroll through the garden / Audrey Terras.
p. cm. – (Cambridge studies in advanced mathematics ; 128)
ISBN 978-0-521-11367-0 (Hardback)
1. Graph theory. 2. Functions, Zeta. I. Title. II. Series.
QA166.T47 2010
511′.5–dc22

2010024611

ISBN 978-0-521-11367-0 Hardback

Contents

List of illustrations — *page* viii
Preface — xi

Part I A quick look at various zeta functions — 1

1 Riemann zeta function and other zetas from number theory — 3

2 Ihara zeta function — 10
- 2.1 The usual hypotheses and some definitions — 10
- 2.2 Primes in X — 11
- 2.3 Ihara zeta function — 12
- 2.4 Fundamental group of a graph and its connection with primes — 13
- 2.5 Ihara determinant formula — 17
- 2.6 Covering graphs — 20
- 2.7 Graph theory prime number theorem — 21

3 Selberg zeta function — 22

4 Ruelle zeta function — 27

5 Chaos — 31

Part II Ihara zeta function and the graph theory prime number theorem — 43

6 Ihara zeta function of a weighted graph — 45

7 Regular graphs, location of poles of the Ihara zeta, functional equations 47

8 Irregular graphs: what is the Riemann hypothesis? 52

9 Discussion of regular Ramanujan graphs 61
- 9.1 Random walks on regular graphs 61
- 9.2 Examples: the Paley graph, two-dimensional Euclidean graphs, and the graphs of Lubotzky, Phillips, and Sarnak 63
- 9.3 Why the Ramanujan bound is best possible (Alon and Boppana theorem) 68
- 9.4 Why are Ramanujan graphs good expanders? 70
- 9.5 Why do Ramanujan graphs have small diameters? 73

10 Graph theory prime number theorem 75
- 10.1 Which graph properties are determined by the Ihara zeta? 78

Part III Edge and path zeta functions 81

11 Edge zeta functions 83
- 11.1 Definitions and Bass's proof of the Ihara three-term determinant formula 83
- 11.2 Properties of W_1 and a proof of the theorem of Kotani and Sunada 90

12 Path zeta functions 98

Part IV Finite unramified Galois coverings of connected graphs 103

13 Finite unramified coverings and Galois groups 105
- 13.1 Definitions 105
- 13.2 Examples of coverings 111
- 13.3 Some ramification experiments 115

14 Fundamental theorem of Galois theory 117

15 Behavior of primes in coverings 128

16 Frobenius automorphisms 133

17 How to construct intermediate coverings using the Frobenius automorphism 141

18 Artin L-functions 144
18.1 Brief survey on representations of finite groups 144
18.2 Definition of the Artin–Ihara L-function 148
18.3 Properties of Artin–Ihara L-functions 154
18.4 Examples of factorizations of Artin–Ihara L-functions 157

19 Edge Artin L-functions 164
19.1 Definition and properties of edge Artin L-functions 164
19.2 Proofs of determinant formulas for edge Artin L-functions 169
19.3 Proof of the induction property 173

20 Path Artin L-functions 178
20.1 Definition and properties of path Artin L-functions 178
20.2 Induction property 180

21 Non-isomorphic regular graphs without loops or multiedges having the same Ihara zeta function 186

22 Chebotarev density theorem 194

23 Siegel poles 200
23.1 Summary of Siegel pole results 200
23.2 Proof of Theorems 23.3 and 23.5 202
23.3 General case; inflation and deflation 206

Part V Last look at the garden 209

24 An application to error-correcting codes 211

25 Explicit formulas 216

26 Again chaos 218

27 Final research problems 227

References 230
Index 236

Illustrations

1.1	Graph of the modulus of the Riemann zeta	5
1.2	Facts about zeta functions and L-functions	6
1.3	What zeta and L-functions say about number fields	6
1.4	Statistics of prime ideals and zeros	7
1.5	Splitting of primes in quadratic extensions	8
2.1	"Bad" graph for the theory of zeta functions	11
2.2	An arbitrary orientation of the edges of a graph	11
2.3	Bouquet of loops	14
2.4	Part of the 4-regular tree	16
2.5	Tetrahedron graph K_4 and $K_4 - e$	16
2.6	Contour map of the modulus of the reciprocal of Ihara zeta for K_4 in the u-variable	19
2.7	Contour map of the modulus of reciprocal of Ihara zeta for K_4 in the s-variable	19
2.8	The cube as a quadratic covering of the tetrahedron	21
3.1	Failure of Euclid's fifth postulate	23
3.2	Fundamental domain for H mod SL$(2, \mathbb{Z})$	23
3.3	Tessellation of upper half plane from the modular group	24
3.4	Images of points on two geodesic circles after mapping into fundamental domain then by Cayley transform into the unit disc	25
5.1	Spectra and level spacings	32
5.2	Spectra of 200 random real 50×50 symmetric matrices	33
5.3	Spectrum of random normal real 1001×1001 matrix	35
5.4	Level spacing histograms for ^{166}Er and a nuclear data ensemble	36
5.5	Odlyzko's comparison of the level spacings of zeros of the Riemann zeta function and those for the GUE	37

5.6	Distributions of the eigenvalues of an adjacency matrix and zeta poles	41
7.1	Possible poles of $\zeta_X(u)$ for a regular graph	49
8.1	Poles of Ihara zeta function of $X = Y_5$	55
8.2	Poles of Ihara zeta function of a random graph	56
8.3	Poles of Ihara zeta function of a torus graph minus some edges	57
8.4	Mathematica experiment	58
8.5	Mathematica experiment	59
9.1	Part of the proof of Theorem 9.8	70
11.1	The dumbbell graph	87
11.2	Matlab experiment	88
11.3	The paths in Lemma 11.11	92
12.1	Labeling the edges of the tetrahedron	101
13.1	A directed graph and a neighborhood of a vertex	106
13.2	An illegal covering map	107
13.3	A 3-sheeted covering	107
13.4	The cube as a normal quadratic covering of the tetrahedron	108
13.5	A non-normal cubic (3-sheeted) covering of the tetrahedron	109
13.6	An n-cycle as a Galois cover of a loop	112
13.7	Normal Klein 4-group covering Z of the dumbbell X	113
13.8	An order-4 cyclic cover Y/X, where Y is the cube	114
13.9	A Klein 4-group cover Y/X, where Y is the cube	114
13.10	A cyclic 6-fold cover Y/X, where Y is the octahedron	115
13.11	Edge ramified cover of K_4	116
14.1	A covering isomorphism of intermediate graphs	118
14.2	Part of the proof of part (4) of Theorem 14.3	120
14.3	Part of the proof of Theorem 14.5	123
14.4	A 6-sheeted normal cover with intermediate covers	125
15.1	Example of the splitting of unramified primes in a non-normal cubic extension of the rationals	129
15.2	Splitting of prime with $f = 2$	130
15.3	Prime which splits completely ($f = 1$)	131
15.4	Splitting of primes in a non-normal cubic cover	132
16.1	Definition of Frobenius symbol and Artin L-function of Galois extension of number fields	134
16.2	Applications of Artin L-functions of number fields	135
16.3	Frobenius automorphism and the normalized version	136
16.4	The map σ preserves the composition of paths	136
16.5	Part of the proof of Proposition 16.5	138
19.1	Edge labelings for the cube as a $\mathbb{Z}_4$ covering of the dumbbell	167

19.2 Proving the induced representation property of edge L-functions 176
20.1 A 12-cyclic cover of the base graph with two loops and two vertices 179
20.2 Eigenvalues of the edge adjacency matrix for a large cyclic cover 181
20.3 The contraction of sheets of a cover corresponding to the contraction of the spanning tree 182
20.4 Contracted versions of X and Y_3 from Figure 14.4 183
21.1 Buser's isospectral non-isomorphic Schreier graphs 187
21.2 Non-isomorphic graphs without loops or multiedges having the same Ihara zeta function 188
21.3 Two isospectral non-isomorphic graphs, named Harold and Audrey 192
22.1 Chebotarev density theorem in the number field case 195
23.1 Paths in the proof of Theorem 23.3 for the case when the nodes are different 203
23.2 The covering appearing in Theorem 23.5 205
23.3 An inflation of X increasing the length of paths by a factor 3 206
24.1 A Tanner graph and a quadratic cover 212
24.2 The normal graph corresponding to a Tanner graph 213
26.1 Matlab experiment showing the spectrum of a random 2000×2000 W_1-type matrix 219
26.2 Normalized nearest neighbor spacing for the spectrum of the matrix in Figure 26.1 220
26.3 Histogram of nearest neighbor spacings of the spectrum of the random graph from Figure 11.2 and the modified Wigner surmise from formula (26.2) with $\omega = 3$ and 6 221
26.4 Tom Petrillo's figure showing the region bounding the spectrum of an edge adjacency operator 222
26.5 Random 3-cover of two loops with an extra vertex 223
26.6 Matlab experiment with spectrum W_1 for large random cover 224
26.7 Nearest neighbor spacings for the spectrum of the edge adjacency matrix of the previous figure 225
26.8 Matlab experiment showing the eigenvalues of the edge adjacency matrix for a large abelian cover 225
26.9 Histogram of the nearest neighbor spacings for the spectrum in Figure 26.8 226

Preface

The goal of this book is to guide the reader in a stroll through the garden of zeta functions of graphs. The subject arose in the late part of the twentieth century and was modelled on the zetas found in other gardens.

Number theory involves many zetas, starting with Riemann's – a necessary ingredient in the study of the distribution of prime numbers. Other zetas of interest to number theorists include the Dedekind zeta function of an algebraic number field and its analog for function fields. Many Riemann hypotheses have been formulated and a few proved. The statistics of the complex zeros of zeta have been connected with the statistics of the eigenvalues of random

Hermitian matrices (the Gaussian unitary ensemble (GUE) distribution of quantum chaos). Artin L-functions are also a kind of zeta associated with a representation of a Galois group of number or function fields. We will find graph analogs of all of these.

Differential geometry has its own zeta, the Selberg zeta function, which is used to study the distribution of the lengths of prime geodesics in compact or arithmetic Riemann surfaces. There is a third zeta function, known as the Ruelle zeta function, which is associated with dynamical systems. We will look at these zetas briefly in Part I. The graph theory zetas are related to these zetas too.

In Part I we give a brief glimpse of four sorts of zeta function, to motivate the rest of the book. In fact, much of Part I is not necessary for the rest of the book. Feel free to skip all but Chapter 2 on the Ihara zeta function.

Prerequisites for reading this book include linear algebra and group theory. What do groups have to say about graphs which appear to have no symmetry? The answer comes with an understanding of the fundamental group whose elements are closed paths through a vertex. This group is intrinsic to our subject. We will find that the theory Galois developed at a young age has its applications here. Our zetas are reciprocals of polynomials, sometimes in several variables. We will obtain determinant formulas for these zetas. And Galois theory will lead to factorizations of the zetas of normal covering graphs, just as it leads to factorizations of Dedekind zeta functions of Galois extensions of number fields.

Most of this book arises from joint work with Harold Stark. Thanks are due to the many people who listened to my lectures on this book and helped with the research, Matthew Horton, Derek Newland, Tom Petrillo, Adriano Garsia, Angela Hicks, Paul Horn, and Yeon Kyung Kim at the University of California, San Diego. I would also like to thank the people who encouraged me by attending my Ulam Seminar at the University of Colorado, Boulder, especially Lynne Walling, David Grant, Su-ion Ih, Vinod Radhakrishnan, Erika Frugoni, Jonathan Kish, and Mike Daniel.

In addition, I want to thank the Newton Institute in Cambridge, England for its hospitality while I worked on this manuscript, which was prepared using Scientific Workplace.

Part I

A quick look at various zeta functions

In Part I we give a brief introduction to the zeta functions of Riemann, Ihara, Selberg, and Ruelle. This part ends with a look at quantum chaos and random matrix theory.

1

Riemann zeta function and other zetas from number theory

There are many popular books about the Riemann zeta and many "serious" ones as well. Serious references for this topic include Davenport [34], Edwards [37], Iwaniec and Kowalski [64], Miller and Takloo-Bighash [86], and Patterson [97]. I googled "zeta functions" today and got around 181 000 hits. The most extensive website was www.aimath.org.

The theory of zeta functions was developed by many people but Riemann's work in 1859 was certainly the most important. The concept was generalized for the purposes of number theorists by Dedekind, Dirichlet, Hecke, Takagi, Artin, and others. Here we will concentrate on the original, namely Riemann's zeta function. The definition is as follows.

Riemann's zeta function for $s \in \mathbb{C}$ with $\operatorname{Re} s > 1$ is defined to be

$$\zeta(s) = \sum_{n=1}^{\infty} \frac{1}{n^s} = \prod_{p \text{ prime}} \left(1 - \frac{1}{p^s}\right)^{-1}.$$

The infinite product here is called an **Euler product**. In 1859 Riemann extended the definition of zeta to a function that is analytic in the whole complex plane except that it has a simple pole at $s = 1$. He also showed that there is an unexpected symmetry known as the **functional equation** relating the value of zeta at s and the value at $1 - s$. It says

$$\Lambda(s) \equiv \pi^{-s/2} \Gamma\left(\frac{s}{2}\right) \zeta(s) = \Lambda(1 - s). \tag{1.1}$$

The **Riemann hypothesis** (**RH**) says that the non-real zeros of $\zeta(s)$ (equivalently those with $0 < \operatorname{Re} s < 1$) are on the line $\operatorname{Re} s = 1/2$. It is equivalent to giving the best possible error term in the prime number theorem in formula (1.2) below. The Riemann hypothesis was checked to 10^{13}th zero (October 12,

2004) by Xavier Gourdon with the help of Patrick Demichel. See Ed Pegg Jr's website for an article called the "Ten trillion zeta zeros":

http://www.maa.org/editorial/mathgames

You win \$1 million if you have a proof of the Riemann hypothesis. See the Clay Mathematics Institute website:

www.claymath.org

A. Odlyzko has studied the spacings of the zeros and found that they appear to be the spacings of the eigenvalues of a random Hermitian matrix (a Gaussian unitary ensemble (GUE)). See Figure 5.5 and the paper on Odlyzko's website

www.dtc.umn.edu/~odlyzko/doc/zeta.htm

If one knows the Hadamard product formula for zeta (from a graduate complex analysis course) as well as the Euler product formula (1.1) above, one can obtain explicit formulas displaying a relationship between primes and the zeros of zeta. Such reasoning ultimately led Hadamard and de la Vallée Poussin to prove the prime number theorem, about 50 years after Riemann's paper. The **prime number theorem** says

$$\#\{p = \text{prime} | p \leq x\} \sim \frac{x}{\log x} \qquad \text{as } x \to \infty. \tag{1.2}$$

Figure 1.1 is a graph of $z = |\zeta(x + iy)|$ drawn using Mathematica. The cover of *The Mathematical Intelligencer*, vol. 8, no. 4, 1986, shows a similar graph with the pole at $x + iy = 1$ and the first six zeros, which are on the line $x = 1/2$ of course. The picture was made by D. Asimov and S. Wagon to accompany their article on the evidence for the Riemann hypothesis. The Mathematica people will sell you a huge poster of the Riemann zeta function.

Exercise 1.1 Use Mathematica (or your favorite software) to do a contour plot of the Riemann zeta function in the same region as that of Figure 1.1.

Hint: Mathematica has a command to give you the Riemann zeta function. It is `Zeta[s]`.

The explicit formulas mentioned above say that sums over the zeros of the zeta function are equal to sums over the primes. References are Murty [91] and Miller and Takloo-Bighash [86].

Many other kinds of zeta function have been investigated since Riemann. In number theory there is the **Dedekind zeta function** of an algebraic number field K, such as $K = \mathbb{Q}(\sqrt{2})$, for example. This zeta is an infinite product

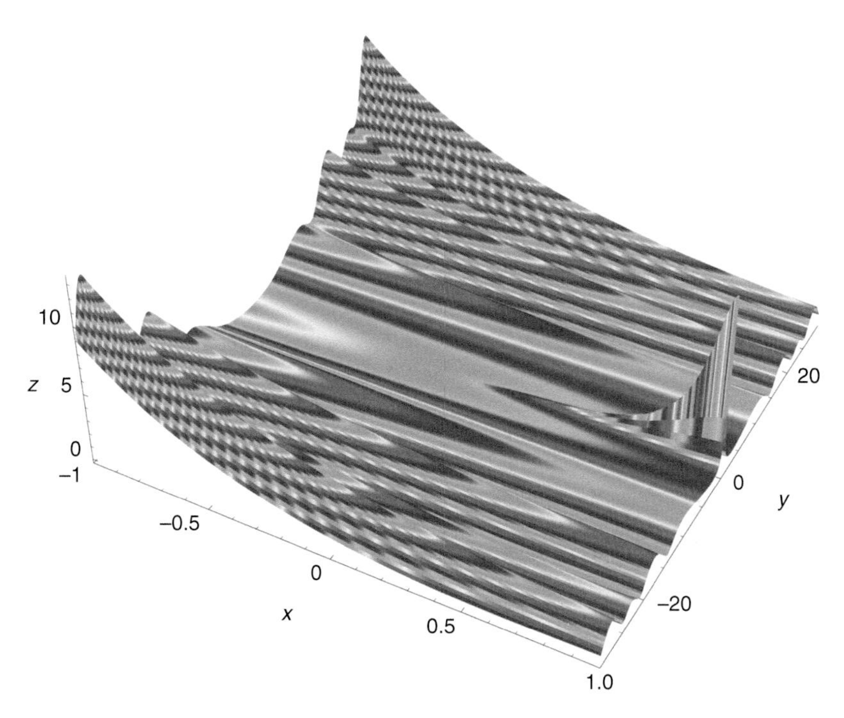

Figure 1.1 Graph of the modulus of the Riemann zeta, i.e., $z = |\zeta(x + iy)|$, showing the pole at $x + iy = 1$ and the complex zeros nearest the real axis (all of which are on the line Re $s = 1/2$, of course).

over prime ideals $\mathfrak{p}$ in O_K, the ring of algebraic integers of K. For our example, $O_K = \mathbb{Z}[\sqrt{2}] = \{a + b\sqrt{2} | a, b \in \mathbb{Z}\}$. The terms in the product are $(1 - N\mathfrak{p}^{-s})^{-1}$, where $N\mathfrak{p} = \#(O_K/\mathfrak{p})$. Riemann's work can be extended to this zeta function and it can be used to prove the prime ideal theorem. The RH is unproved but conjectured to be true for the Dedekind zeta function. Surprisingly, no one has yet proved (even in the case of quadratic number fields, $K = \mathbb{Q}(\sqrt{m})$), where m is a non-square integer, that there cannot be a real zero near 1. Such a possible zero is called a "**Siegel zero**." A reference for this zeta is Lang's book [73], where it is explained why the non-existence of Siegel zeros would lead to many nice consequences for number theory. Figures 1.2–1.5 give summaries of the basic facts about zeta and L-functions for $\mathbb{Q}$ and $\mathbb{Q}(\sqrt{d})$. We will find graph theory analogs of many of these facts.

There are also **function field zeta functions**, where the number field K is replaced by a finite algebraic extension of $\mathbb{F}_q(x)$, the rational functions of one variable over the finite field $\mathbb{F}_q$ with q elements. André Weil proved the RH for this zeta, which is a rational function of $u = q^{-s}$. See Rosen [104].

Dedekind zeta	$\zeta_K(s) = \prod_{\mathfrak{p}}(1 - N\mathfrak{p}^{-1})^{-s}$
	product over prime ideals in O_K, $N\mathfrak{p} = \#(O_K/\mathfrak{p})$
Riemann zeta for F = $\mathbb{Q}$	$\zeta_{\mathbb{Q}}(s) = \prod_p (1 - p^{-s})^{-1}$
	product over primes in $\mathbb{Z}$
Dirichlet L-function	$L(s, \chi) = \prod_p (1 - \chi(p) Np^{-s})^{-1}$
	$\chi(p) = (2/p)$,
	product over primes in $\mathbb{Z}$
Factorization	$\zeta_{\mathbb{Q}(\sqrt{2})}(s) = \zeta_{\mathbb{Q}}(s) L(s, \chi)$

Figure 1.2 A summary of facts about the zeta functions and the L-functions associated with the number fields $\mathbb{Q}$ and $\mathbb{Q}(\sqrt{2})$. See Figure 1.5 for a definition of the Legendre symbol $(2/p)$.

Functional equations: $\zeta_K(s)$ related to $\zeta_K(1-s)$ (Hecke)

values at 0: $r = r_1 + r_2 - 1$; r_1, number of real conjugate fields of K over $\mathbb{Q}$; r_2, number of pairs of complex conjugate fields of K over $\mathbb{Q}$. If $K = \mathbb{Q}(\sqrt{2})$ then $r_1 = 2$, $r_2 = 0$.

$$\zeta(0) = -\frac{1}{2}, \quad [s^{-r}\,\zeta_K(s)]\Big|_{s=0} = \frac{-hR}{w}$$

h, class number, measures how far O_K is from having unique factorization; $h = 1$ for $K = \mathbb{Q}(\sqrt{2})$

R, regulator (determinant of logs of units)

$$R = \log(1 + \sqrt{2}) \textbf{ when } K = \mathbb{Q}(\sqrt{2})$$

w, number of roots of unity in K is 2, when $K = \mathbb{Q}(\sqrt{2})$

Figure 1.3 What the zeta and L-functions say about the number fields.

Statistics of prime ideals and zeros

- **From information on zeros of $\zeta_K(s)$ obtain prime ideal theorem in number fields**

$$\#\{\mathfrak{p} \text{ prime ideal in } O_K | N\mathfrak{p} \leq x\} \sim \frac{x}{\log x} \quad as\ x \to \infty$$

- **There are an infinite number of primes p such that $\left(\frac{2}{p}\right) = 1$.**

- **Dirichlet theorem: there are an infinite number of primes p in the progression**

  $a,\ a+d,\ a+2d,\ a+3d,\ \ldots$ **when** g.c.d.$(a,d) = 1$.

- **Riemann hypothesis is still open for number fields; done for function fields by André Weil:**
  GRH or ERH: $\zeta_K(s) = 0$ implies $\operatorname{Re} s = 1/2$, assuming s is not real.

Figure 1.4 Statistics of prime ideals and zeros: g.c.d., greatest common divisor; GRH, generalized Riemann hypothesis; ERH, extended Riemann hypothesis.

Another generalization of Riemann's zeta function is the Dirichlet L-function associated with a multiplicative character χ defined on the group of integers $a \pmod m$ with a relatively prime to m. This function is thought of as a function on the integers which is 0 unless a and m have no common divisors. Then one has the **Dirichlet L-function**, for $\operatorname{Re} s > 1$ defined by

$$L(s, \chi) = \sum_{n=1}^{\infty} \frac{\chi(n)}{n^s}.$$

This L-function also has an Euler product, analytic continuation, functional equation, Riemann hypothesis (the extended Riemann hypothesis or ERH). This function can be used to prove the Dirichlet theorem stating that there are infinitely many primes in an arithmetic progression of the form $a, a+d, a+2d, a+3d, \ldots, a+kd, \ldots$, assuming that a and d are relatively prime. More generally there are **Artin L-functions** attached to representations of Galois groups of normal extensions of number fields. The Artin conjecture,

Quadratic extension

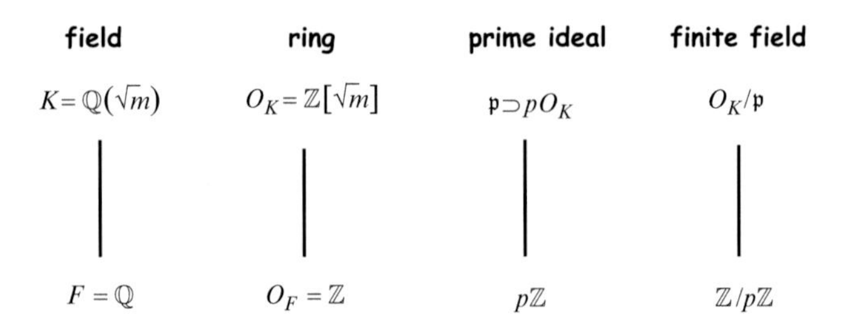

g, **# of such** $\mathfrak{p}$**;** f, **degree of** $O_K/\mathfrak{p}$ **over** O_F/pO_F**;** $efg = 2$

Assume that m **is a square-free integer congruent to** 2 **or** 3 (mod 4).

Decomposition of primes in quadratic extensions

$$K = F(\sqrt{m})/F, \qquad F = \mathbb{Q}$$

Three cases:

(1) p **inert,** $f = 2$: $pO_K =$ **prime ideal in** K, $m \not\equiv x^2 \pmod{p}$

(2) p **splits,** $g = 2$: $pO_K = \mathfrak{p}\mathfrak{p}'$, $\mathfrak{p} \neq \mathfrak{p}'$, $m \equiv x^2 \pmod{p}$

(3) p **ramifies,** $e = 2$: $pO_K = \mathfrak{p}^2$, p **divides** $4m$

$\mathrm{Gal}(K/F) = \{1, -1\}$

Frobenius automorphism (Legendre symbol)

$$\left(\frac{4m}{p}\right) = \begin{cases} -1 & \text{in case (1)} \\ 1 & \text{in case (2)} \\ 0 & \text{in case (3)} \end{cases}$$

If p **does not divide** $4m$ **then** p **has 50% chance of being in case (1) and 50% chance of being in case (2).**

Assume that m **is a square-free integer equal to** 2 **or** 3 (mod 4).

Figure 1.5 Splitting of primes in quadratic extensions. *Top, moving left to right,* the four vertical lines represent respectively the number field extension $\mathbb{Q}(\sqrt{m})/\mathbb{Q}$, the corresponding rings of integers, the prime ideals, and the finite residue fields. Here f is the degree of the extension of finite residue fields, g is the number of primes of O_K containing the prime p of $\mathbb{Z}$, and e is the ramification exponent. We have $efg = 2$ in the present case, for which $K = \mathbb{Q}(\sqrt{m})$.

as yet unproved, says that if the representation is irreducible and not trivial (i.e., not identically 1), the L-function is entire. These L-functions were named for Emil Artin. A reference for Artin L-functions is Lang [73]. We will be interested in graph theory analogs of Artin L-functions.

Yet another sort of zeta is the **Epstein zeta function** attached to a quadratic form

$$Q[x] = \sum_{i,j=1}^{n} q_{ij} x_i x_j.$$

We will assume that the q_{ij} are real and that Q is positive definite, meaning that $Q[x] > 0$, if $x \neq 0$. Then the **Epstein zeta function** is defined for complex s with $\mathrm{Re}\, s > n/2$ by:

$$Z(Q, s) = \sum_{a \in \mathbb{Z}^n - 0} Q[a]^{-s}.$$

As in the case of the Riemann zeta, there is an analytic continuation to all $s \in \mathbb{C}$ with a pole at $s = n/2$. And there is a functional equation relating $Z(Q, s)$ and $Z(Q, n - s)$. Even when $n = 2$, the analog of the Riemann hypothesis may be false for the Epstein zeta function. See Terras [132] for more information on this zeta function.

If $Q[x] \in \mathbb{Z}$ for all $x \in \mathbb{Z}^n$ then, defining $N_m(Q) = \left|\{x \in \mathbb{Z}^n | Q[x] = m\}\right|$, we see that $Z(Q, s) = \sum_{m \geq 1} N_m m^{-s}$, assuming $\mathrm{Re}\, s > n/2$. Similarly, one can define zeta functions attached to many lists of numbers such as $N_m(Q)$, in particular to the Fourier coefficients of modular forms. Classically modular forms are holomorphic functions on the upper half plane having an invariance property under a group of fractional linear transformations such as the modular group SL(2, $\mathbb{Z}$) consisting of 2×2 matrices with integer entries and determinant 1. See Miller and Takloo-Bighash [86], Sarnak [109], or Terras [132] for more information. Now the idea of modular forms has been vastly generalized and even plays a role in Andrew Wiles' proof of Fermat's last theorem.

2
Ihara zeta function

2.1 The usual hypotheses and some definitions

Our graphs will be finite, connected, and undirected. It will usually be assumed that they contain no degree-1 vertices, called "leaves" or "hair" or "danglers". We will also usually assume that the graphs are not cycles or cycles with hair. A **cycle graph** is obtained by arranging the vertices in a circle and connecting each vertex to the two vertices next to it on the circle. A "bad" graph – meaning that it does not satisfy the above assumptions – is pictured in Figure 2.1. We will allow our graphs to have loops and multiple edges between pairs of vertices.

Why do we make these assumptions? They are necessary hypotheses for many of the main theorems (for example, the graph theory prime number theorem, formula (2.4)). References for graph theory include Biggs [15], Bollobás [19], Fan Chung [26], and Cvetković, Doob, and Sachs [32].

A **regular graph** is a graph each of whose vertices has the same **degree**, i.e., the same number of edges coming out of the vertex. A graph is k-**regular** if every vertex has degree k. **Simple graphs** have no loops or multiple edges. Our graphs need not be regular or simple. A **complete graph** K_n on n vertices has all possible edges between its vertices but no loops.

Definition 2.1 Let V denote the vertex set of a graph X with $n = |V|$. The **adjacency matrix** A of X is an $n \times n$ matrix with (i, j)th entry

$$a_{ij} = \begin{cases} \text{number of undirected edges connecting vertex } i \text{ to vertex } j, & \text{if } i \neq j; \\ 2 \times \text{number of loops at vertex } i, & \text{if } i = j. \end{cases}$$

In order to define the Ihara zeta function, we need to define a prime in a graph X with edge set E having $m = |E|$ elements. To do this, we first direct

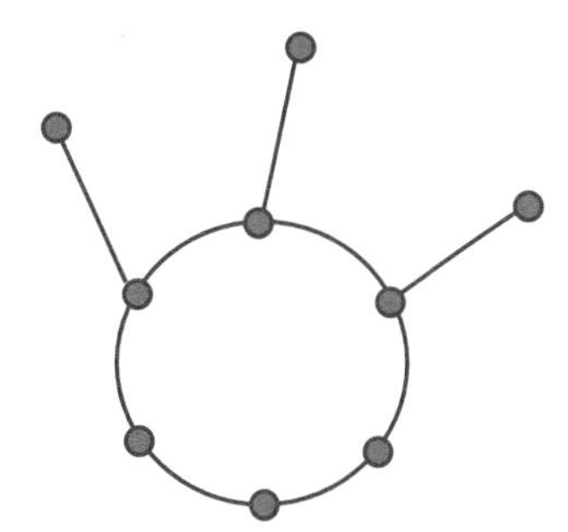

Figure 2.1 This is an example of a "bad" graph for the theory of zeta functions. For this graph, there are only finitely many primes (two to be exact), as defined below.

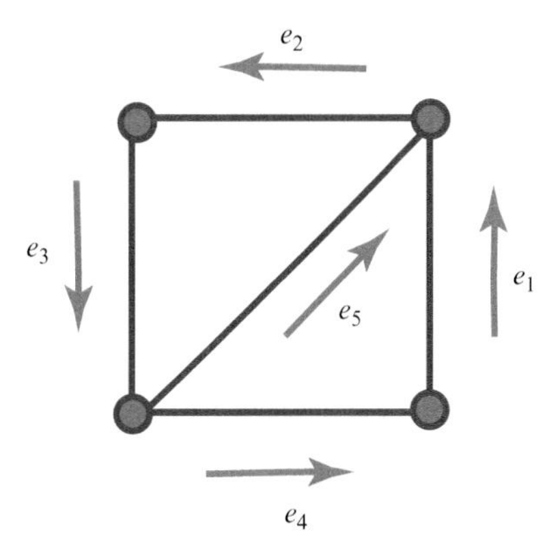

Figure 2.2 We choose an arbitrary orientation of the edges of a graph. Then we label the inverse edges (edges traveled in the opposite direction) by $e_{j+5} = e_j^{-1}$ for $j = 1, \ldots, 5$.

or orient the edges of our graph arbitrarily and **label the edges** as follows:

$$e_1, \quad \ldots, \quad e_m, \quad e_{m+1} = e_1^{-1}, \quad \ldots, \quad e_{2m} = e_m^{-1}. \qquad (2.1)$$

Here $m = |E|$ is the number of unoriented edges of X and $e_j^{-1} = e_{j+m}$ is the edge e_j with the opposite orientation. See Figure 2.2 for an example.

2.2 Primes in X

A **path** or walk $C = a_1 \cdots a_s$, where a_j is an oriented edge of X, is said to have a **backtrack** if $a_{j+1} = a_j^{-1}$ for some $j = 1, \ldots, s-1$. A path $C = a_1 \cdots a_s$ is said to have a **tail** if $a_s = a_1^{-1}$. The **length** of $C = a_1 \cdots a_s$ is $s = \nu(C)$. A **closed path** or **cycle** means that the starting vertex is the same as the terminal vertex. The closed path $C = a_1 \cdots a_s$ is called a **primitive or prime path** if it has no backtrack or tail and $C \neq D^f$ for $f > 1$. That is, you can only go around the path once. For the closed path $C = a_1 \cdots a_s$, the **equivalence class** $[C]$ means the following

$$[C] = \{a_1 \cdots a_s, \; a_2 \cdots a_s a_1, \; \ldots, \; a_s a_1 \cdots a_{s-1}\}. \tag{2.2}$$

That is, we call two closed paths **equivalent** if we can get one from the other by changing the starting vertex. A **prime** in the graph X is an equivalence class $[C]$ of prime paths. The **length of the path** C is $\nu(C) = s$, the number of edges in C.

For examples of primes, consider Figure 2.2. We have primes $[C] = [e_2e_3e_5]$, $[D] = [e_1e_2e_3e_4]$, $E = [e_1e_2e_3e_4e_1e_{10}e_4]$. Here $e_{10} = e_5^{-1}$ and the lengths of these primes are given by $\nu(C) = 3$, $\nu(D) = 4$, $\nu(E) = 7$. We have infinitely many primes since $E_n = [(e_1e_2e_3e_4)^n e_1e_{10}e_4]$ is prime for all $n \geq 1$. The only non-primes are powers of primes. In particular, for the graph theory version of things, one does not have unique factorization into primes.

2.3 Ihara zeta function

Definition 2.2 The **Ihara zeta function** for a finite connected graph (without degree-1 vertices) is defined to be the following function of the complex number u, with $|u|$ sufficiently small:

$$\zeta_X(u) = \zeta(u, X) = \prod_{[P]} \left(1 - u^{\nu(P)}\right)^{-1},$$

where the product is over all primes $[P]$ in X. Recall that $\nu(P)$ denotes the length of P.

In the product defining the Ihara zeta function, we distinguish the prime $[P]$ from $[P^{-1}]$, which is the path traversed in the opposite direction. Generally the product is infinite. We will see later how small $|u|$ needs to be for the product to converge. There is one case, however, when the product is finite. Normally we will exclude this case, the cycle graph.

Example 2.3: Cycle graph Let X be a cycle graph with n vertices. Then, since there are only two primes,

$$\zeta_X(u) = (1 - u^n)^{-2}.$$

As a power series in the complex variable u, the Ihara zeta function has non-negative coefficients. Thus, by a classic theorem of Landau, both the series and the product defining $\zeta_X(u)$ will converge absolutely in a circle $|u| < R_X$ with a singularity (a pole of order 1 for connected X) at $u = R_X$. See Apostol [3], p. 237, for Landau's theorem.

Definition 2.4 The **radius of the largest circle of convergence** of the Ihara zeta function is R_X.

In fact, R_X is rather small. When X is a $(q+1)$-regular graph, $R_X = 1/q$. We will say more about the size of R_X for irregular graphs later. Amazingly, the Ihara zeta function is the reciprocal of a polynomial by Theorem 2.5 below.

2.4 Fundamental group of a graph and its connection with primes

One of the favorite determinant formulas for the Ihara zeta function (in Theorem 2.5 below) involves the **fundamental group** $\Gamma = \pi_1(X, v)$ of the graph X. Later we will even define the path zeta, which is more clearly attached to this group.

The fundamental group of a topological space such as our graph X has elements which are closed directed paths starting and ending at a fixed basepoint $v \in X$. Two paths are **equivalent** iff one can be continuously deformed into the other (i.e., one is "homotopic" to the other within X while still starting and ending at vertex v). The **product** of two paths a, b means first go around a then b.

It turns out (by the Seifert–Van Kampen theorem, for example) that the fundamental group of graph X is a free group on r generators, where r is the number of edges left out of a spanning tree for X. Let us explain this further; more information can be found on the web, e.g., in Chapter 1 of the algebraic topology book of Allen Hatcher: www.math.cornell.edu/~hatcher. You could also look at Massey [83], p. 198, or Gross and Tucker [47].

What is a **free group** G on a set S of r generators? Here r is the **rank** of G. The group G is the set of **words** obtained by forming finite strings or words $a_1 \cdots a_t$ of symbols $a_j \in S$ modulo an equivalence relation. Two words $a_1 a_2 \cdots a_t$ and $a_1 a_2 \cdots b b^{-1} \cdots a_t$, $b \in S$, are called equivalent. The product of the words $a_1 \cdots a_t$ and $b_1 \cdots b_s$ is $a_1 \cdots a_t b_1 \cdots b_s$.

What is a spanning tree T in a graph X? First we say that a graph T is a **tree** if it is a connected simple graph without any closed backtrackless paths of length ≥ 1. A **spanning tree** T for graph X means a tree which is a subgraph of X containing all the vertices of X. Every graph has a spanning tree.

Figure 2.3 A bouquet of loops.

From the graph X we construct a new graph $X^{\#}$ by shrinking a spanning tree T of X to a point. The new graph will be a bouquet of r loops as in Figure 2.3. It turns out that the fundamental group of X is the same as that of $X^{\#}$. Why? The quotient map $X \to X/T$ is what algebraic topologists call a "homotopy equivalence." This means that intuitively you can continuously deform one graph into the other. For more information, see Allen Hatcher, Chapter 0 of www.math.cornell.edu/~hatcher.

So what is the fundamental group of the bouquet of r loops in Figure 2.3? We claim it is clearly the free group on r generators. The generators are the loops! The elements are the words in these loops modulo the equivalence relation defined above for words in a free group. The rank r of the fundamental group of the original graph X is thus the number of edges left out of X in order to get a spanning tree.

Exercise 2.1 (a) Find the fundamental groups for the graphs in Figures 2.1, 2.3, and 2.5.

(b) Show that if r is the rank of the fundamental group then $r - 1 = |E| - |V|$.

We have a one-to-one correspondence between **conjugacy classes** $\{C\} = \{xCx^{-1} | x \in \Gamma = \pi_1(X, v)\}$ and equivalence classes of the backtrackless, tailless, cycles $[C^*]$ in X defined in formula (2.2). If a closed path C starting and ending at point v gives rise to a conjugacy class $\{C\}$ in Γ then we may take C in its homotopy class, so that C has no backtracking. We then remove the tail from C so as to get a tailless cycle C^*. The one-to-one correspondence referred to above comes from the fact that the conjugacy class of C in Γ corresponds to the equivalence class of the backtrackless, tailless, cycle C^*. It can be shown (**exercise**) that the change of C in its conjugacy class corresponds to a change of C^* in its equivalence class. In the other direction of the correspondence, given C^* we grow a tail so as to reach v, thus obtaining a path C which determines an element of Γ. A different tail simply conjugates C. Another way of thinking about the correspondence is that the elements of the equivalence class of C^* are precisely the closed cycles of minimal length which are freely homotopic to C. **Freely homotopic** means that the base point v is not fixed.

The fundamental group Γ is a free group of rank r. Thus, for $\gamma \neq 1$ in Γ, the **centralizer** $C_\Gamma(\gamma) = \{\delta \in \Gamma | \gamma\delta = \delta\gamma\}$ is a cyclic subgroup of Γ. Under the one-to-one correspondence between classes $[C]$ of backtrackless tailless cycles in Γ, prime cycles P correspond to conjugacy classes $\{P\}$ in Γ such that the centralizer $C_\Gamma(P)$ is generated by P. Such conjugacy classes $\{P\}$ are called **primitive**. Thus primes $[P]$ in X are in one-to-one correspondence with conjugacy classes of primitive elements in the fundamental group of X. We will say more about this correspondence in Chapter 12 on path zeta functions of graphs.

Note that although an irregular graph may not appear at first glance to have any "symmetry" in the sense of a group of symmetries, there is always the fundamental group lurking around.

We remark here that we cannot always be consistent in our notation. As for as possible we will use italic capital letters for paths in a graph and small Greek letters for elements of the fundamental group (or a Galois group acting on a graph covering), but sometimes conflicts will arise and rigorous consistency will be impossible.

Exercise 2.2 Prove that the centralizer of $\gamma \neq 1$ in the fundamental group Γ is cyclic.

Algebraic topology (see the references above) tells us that there is a "universal covering tree" T (meaning that it is without cycles and is a covering of the original graph X as in Definition 2.9 below). See Figure 2.4 for a picture of part of the 4-regular tree which is the universal cover of any 4-regular graph. Of course, T is infinite.

Exercise 2.3 Draw part of the universal covering tree for the irregular graph $K_4 - e$ obtained by removing one edge from the tetrahedron K_4, the complete graph on four vertices. The graph $K_4 - e$ is pictured on the right in Figure 2.5 and K_4 is on the left.

There is an action of the fundamental group Γ on T such that we can identify T/Γ with X. You can also view the tree in the $(p+1)$-regular case as coming from p-adic matrix groups. See Serre [113], Trimble [137], and the last chapter of Terras [133] as well as [59] and [136].

Figure 2.4 Part of the 4-regular tree.

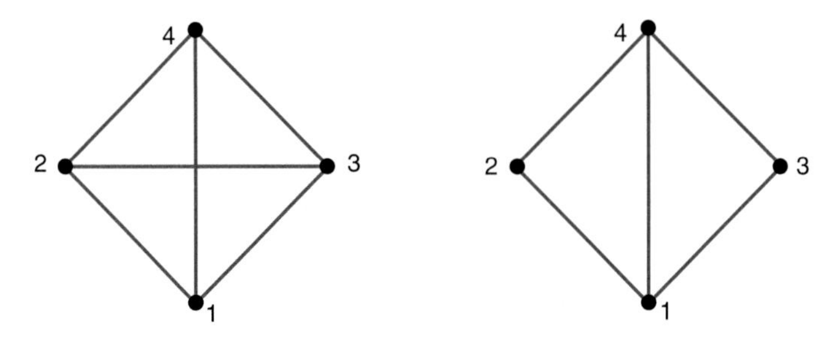

Figure 2.5 *Left*, the tetrahedron graph, also known as K_4, the complete graph on four vertices. *Right*, the graph $K_4 - e$ obtained from the tetrahedron by deleting the edge e joining vertices 2 and 3.

If V is the set of vertices of graph X and E the set of edges then the **Euler characteristic** of X is $\chi(X) = |V| - |E| = 1 - r$, where r is the rank of the fundamental group Γ of X. See Exercise 2.1.

One moral of the preceding considerations is that we can rewrite the product in Definition 2.2 in the language of the fundamental group of X as

$$\zeta_X(u) = \prod_{\{C\}} \left(1 - u^{\nu(C)}\right)^{-1}, \tag{2.3}$$

where the product is over primitive conjugacy classes in the fundamental group Γ of X. A primitive conjugacy class $\{C\}$ in Γ means that C generates the centralizer of C in Γ. Here $\nu(C)$ means the length of the element C^* in the equivalence class of tailless backtrackless paths corresponding to $\{C\}$. Alternatively, $\nu(C)$ is the minimal length of all cycles freely homotopic to C. As above, we distinguish between C and C^{-1} in this product.

2.5 Ihara determinant formula

Mercifully, the zeta function can be computed from the Ihara determinant formula in the theorem which follows.

Theorem 2.5 (Ihara theorem generalized by Bass, Hashimoto, etc.) *Let A be the adjacency matrix of X and Q the diagonal matrix with jth diagonal entry q_j such that $q_j + 1$ is the degree of the jth vertex of X. Suppose that r is the rank of the fundamental group of X; $r - 1 = |E| - |V|$. Then we have the Ihara three-term determinant formula*

$$\zeta_X(u)^{-1} = \left(1 - u^2\right)^{r-1} \det\left(I - Au + Qu^2\right).$$

Exercise 2.4 Show that $r - 1 = \frac{1}{2}\operatorname{Tr}(Q - I)$.

There is an elementary proof of the preceding theorem using the method of Bass [12]. We will present it in Part III. In [133] we presented another proof for k-regular graphs using the Selberg trace formula on the k-regular tree. For k-regular graphs, the Ihara zeta function has much in common with the Riemann zeta function.

Suppose that X is a $(q + 1)$-regular graph. Then the Ihara zeta function has functional equations relating the value at u to the value at $1/(qu)$. Setting $u = q^{-s}$, one finds that there is a functional equation relating the value at s to that at $1 - s$, just as for Riemann's zeta function. See Proposition 7.5 below.

When the graph X is $(q + 1)$-regular, there is also an analog of the Riemann hypothesis. It turns out to hold if and only if the graph is Ramanujan as defined by Lubotzky *et al.* in [79]. See Definition 7.3 and Theorem 7.4.

Definition 2.6 A connected $(q + 1)$-regular graph X is **Ramanujan** if and only if, for

$$\mu = \max\left\{|\lambda| \middle| \lambda \in \text{Spectrum } A,\ |\lambda| \neq q + 1\right\}.$$

we have $\mu \leq 2\sqrt{q}$.

Some graphs are Ramanujan and some are not. In the 1980s, Margulies and independently Lubotzky *et al.* [79] found a construction of infinite families of Ramanujan graphs of fixed degree equal to $1 + p^e$, where p is a prime. They used the Ramanujan conjecture (now proved by Deligne) to show that the graphs were Ramanujan. Such graphs are of interest to computer scientists because they provide efficient communication as they have good expansion properties. See Giuliana Davidoff *et al.* [35], Lubotzky [77], Sarnak [109], or Terras [132] for more information. Friedman [42] proved that a random regular graph is almost Ramanujan (the **Alon conjecture**).

See Miller *et al.* [87] for experiments leading to the conjecture that the percentage of regular graphs exactly satisfying the RH approaches 27% as the number of vertices approaches infinity. The argument involves the Tracy–Widom distribution from random matrix theory. A useful survey on expander graphs and their applications is that of Hoory, Linial, and Wigderson [55]. We discuss this in Chapter 9.

Example 2.7 The **tetrahedron graph** K_4 in Figure 2.5 has Ihara zeta function

$$\zeta_{K_4}(u)^{-1} = \left(1-u^2\right)^2\left(1-u\right)\left(1-2u\right)\left(1+u+2u^2\right)^3.$$

The five poles of this zeta function are located at the points $-1, 1/2, 1, \left(-1 \pm \sqrt{-7}\right)/4$. The absolute value of the complex pole is $1/\sqrt{2} \cong 0.707\,11$. The closest pole to the origin is $1/2 = 1/q = R_{K_4}$. Of course, K_4 is a Ramanujan graph and thus the Riemann hypothesis holds for this graph.

Next we illustrate what the Ihara zeta counts. The nth coefficient of the generating function $u(d/du)\log\zeta_{K_4}(u)$ is the number N_m of length-n closed paths in K_4 (without backtracking or tails). So there are eight primes of length 3 in K_4, for example. See Definition 4.2 and formula (4.5) proved in Chapter 4, on Ruelle's zeta function. We find that

$$u\frac{d}{du}\log\zeta_{K_4}(u) = 24u^3 + 24u^4 + 96u^6 + 168u^7$$
$$+ 168u^8 + 528u^9 + 1200u^{10} + 1848u^{11} + O\left(u^{12}\right).$$

Figure 2.6 is a contour map of $z = |\zeta_{K_4}(x+iy)|^{-1}$, while Figure 2.7 is a contour map of $z = |\zeta_{K_4}(2^{-(x+iy)})|^{-1}$. The second graph is more like that for the Riemann zeta function in Figure 1.1.

Example 2.8 Let $X = K_4 - e$ be the **graph obtained from the tetrahedron** K_4 **by deleting an edge** e. See Figure 2.5. Then

$$\zeta_X(u)^{-1} = \left(1-u^2\right)\left(1-u\right)\left(1+u^2\right)\left(1+u+2u^2\right)\left(1-u^2-2u^3\right).$$

From this, we have

$$u\frac{d}{du}\log\zeta_X(u) = 12x^3 + 8x^4 + 24x^6 + 28x^7 + 8x^8 + 48x^9 + \cdots.$$

So there are four primes of length 3 in X. There are nine roots: $\pm 1, \pm i, \left(-1 \pm \sqrt{-7}\right)/4$, and three roots of the cubic, s_1, s_2, s_3 with $|s_1| \cong 0.657\,30$, $|s_2| = |s_3| \cong 0.872\,18$. So, for this example $R_X \cong 0.657\,30$. Later we will define what the Riemann hypothesis means for an irregular graph. See Chapter 8.

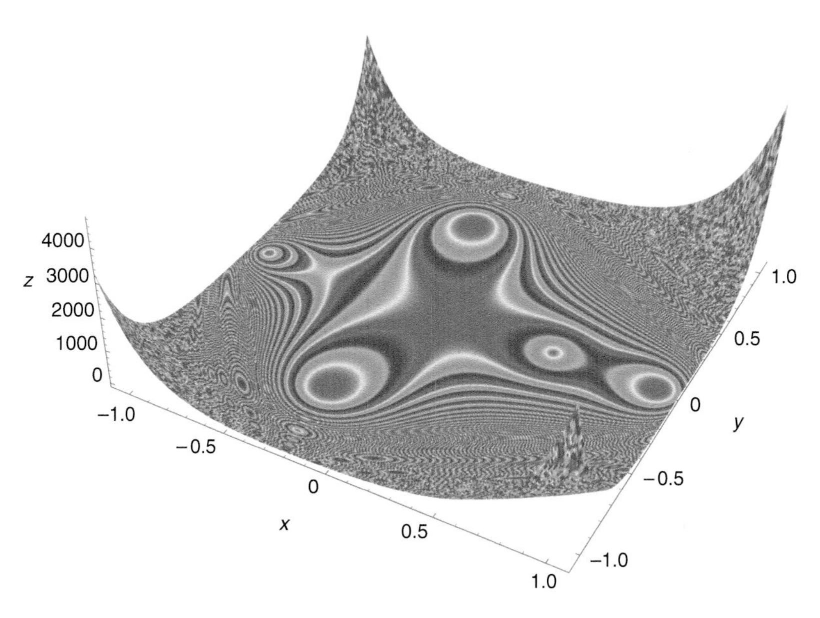

Figure 2.6 A contour map as drawn using Mathematica of the modulus of the reciprocal of the Ihara zeta for K_4 in the u-variable: $z = 1/|\zeta_{K_4}(x + iy)|$. You can see the five roots (not counting multiplicity).

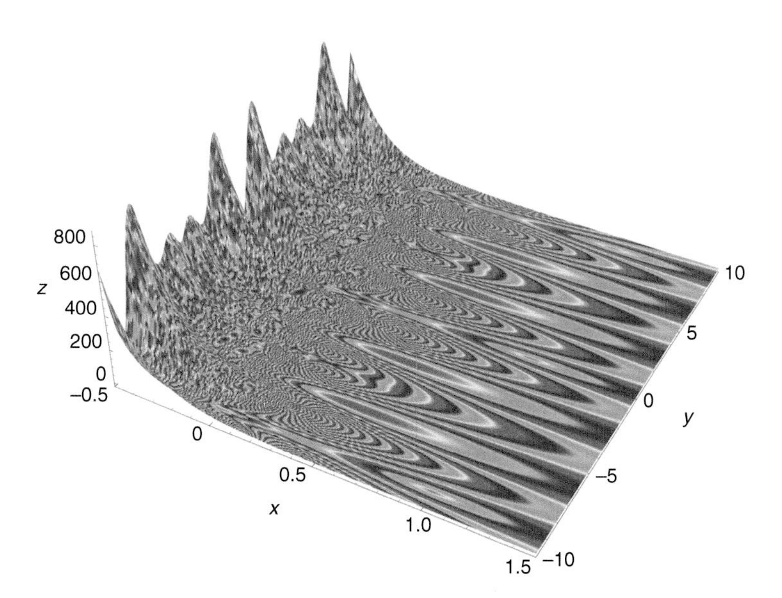

Figure 2.7 A contour map as drawn using Mathematica of the modulus of the reciprocal of zeta for K_4 in the s-variable: $z = 1/|\zeta_{K_4}(2^{-(x+iy)})|$.

The relation between the zeta function for the graph with an edge deleted and the zeta function for the original graph is not obvious. We will have more to say about this topic when we have discussed edge zeta functions.

Exercise 2.5 Compute the Ihara zeta functions of your favorite graphs, e.g., the cube, the dodecahedron, the buckyball.

2.6 Covering graphs

Next we consider an unramified finite covering graph Y of our finite graph X. This is analogous to an extension of algebraic number fields. We assume that both X and Y are connected. A discussion of covering graphs can be found in Massey [83]; see Part IV also. The idea is that locally, at each vertex, the two graphs look alike, though globally they may be very different. In the same way, the sphere and the plane are locally alike.

Definition 2.9 If it has no multiple edges and loops we can say that the graph Y is an **unramified covering** of the graph X if we have a covering map $\pi : Y \to X$ which is an onto graph mapping (i.e., taking adjacent vertices to adjacent vertices) such that, for every $x \in X$ and for every $y \in \pi^{-1}(x)$, the collection of points adjacent to $y \in Y$ is mapped one-to-one onto the collection of points adjacent to $x \in X$.

The factorization of the Ihara zeta function of the quadratic covering in the example below is analogous to what happens for Dedekind zeta functions of quadratic extensions of number fields; see Figure 1.2. In Part IV we will show that the entire theory of Dedekind zeta functions (and Artin L-functions) has a graph theory analog.

Example 2.10: Unramified quadratic covering of K_4 Consider Figure 2.8. The cube Y is obtained by drawing two copies of a spanning tree (whose edges are indicated by dotted lines) for the tetrahedron $X = K_4$ and then drawing the rest of the edges of the cover to go between sheets of the cover. We find that

$$\zeta_Y(u)^{-1} = L\big(u, \rho, Y/X\big)^{-1} \zeta_X(u)^{-1},$$

where

$$L\big(u, \rho, Y/X\big)^{-1} = \big(1 - u^2\big)\big(1 + u\big)\big(1 + 2u\big)\big(1 - u + 2u^2\big)^3.$$

Exercise 2.6 Draw the analogs of Figure 2.6 and 2.7 for the cube.

Exercise 2.7 Find a second quadratic cover Y' of the tetrahedron by drawing two copies of a spanning tree of $X = K_4$ and then connecting the rest of

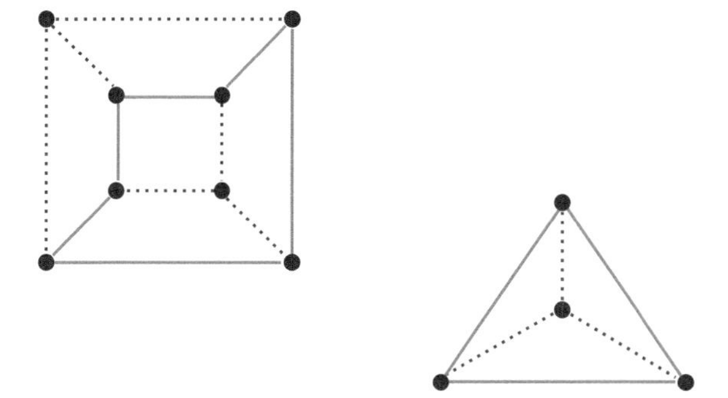

Figure 2.8 The cube is a quadratic covering of the tetrahedron. A spanning tree for the tetrahedron is indicated by the dotted lines. Two copies of this tree are seen in the cube.

the edges of Y so that only two edges in Y' go between sheets of the cover. Compute the Ihara zeta function of Y'.

2.7 Graph theory prime number theorem

One can use the Ihara zeta function to prove the graph prime number theorem. In order to state this result, we need some definitions.

Definition 2.11 The **prime counting function** is

$$\pi(n) = \#\left\{\text{primes } [P] \middle| n = \nu(P) = \text{length of } P\right\}.$$

You may find it a little surprising to note that we have replaced $\leq$ in the usual prime counting function of formula (1.2) with an equals sign here.

Definition 2.12 The **greatest common divisor of the prime path lengths** is

$$\Delta_X = \text{g.c.d.} \left\{\nu(P) \middle| [P] \text{ prime of } X\right\}.$$

The graph theory prime number theorem for a graph X satisfying the usual hypotheses, stated at the beginning of Section 2.1, says that if Δ_X divides m then

$$\pi(m) \sim \frac{\Delta_X}{m R_X^m} \qquad \text{as } m \to \infty. \tag{2.4}$$

If Δ_X does not divide m then $\pi(m) = 0$. Note that the theorem is clearly false for the bad graph in Figure 2.1.

We will see the proof in Chapter 10. It is much easier than the proof of the usual prime number theorem for prime integers. There are also analogs of Dirichlet's theorem on primes in progressions and the Chebotarev density theorem. We will consider this in Chapter 22.

3
Selberg zeta function

Some references for the Selberg zeta function are Dennis Hejhal [52], Atle Selberg [111], [112], Audrey Terras [132], and Marie-France Vignéras [139]. Another reference is the collection of articles edited by Tim Bedford, Michael Keane, and Caroline Series [13].

The Selberg zeta function is a generating function for "primes" in a compact (or finite volume) Riemannian manifold M. Before we define "prime," we need to think a bit about Riemannian geometry. Assuming M has a constant curvature -1, it can be realized as a quotient of the **Poincaré upper half plane**

$$H = \{x + iy | x, y \in \mathbb{R}, y > 0\},$$

with **Poincaré arc length** element

$$ds^2 = \frac{dx^2 + dy^2}{y^2},$$

which can be shown to be invariant under the **fractional linear transformation**

$$z \longrightarrow \frac{az + b}{cz + d}, \qquad \text{where } a, b, c, d \in \mathbb{R} \text{ and } ad - bc > 0.$$

The **Laplace operator** corresponding to the Poincaré arc length is

$$\Delta = y^2 \left(\frac{\partial^2}{\partial x^2} + \frac{\partial^2}{\partial y^2} \right).$$

It too is invariant under fractional linear transformations.

It is not hard to see that **geodesics**, i.e., curves minimizing the Poincaré arc length, are half lines and semicircles orthogonal to the real axis. Calling these geodesics "straight lines" creates a model for non-Euclidean geometry since

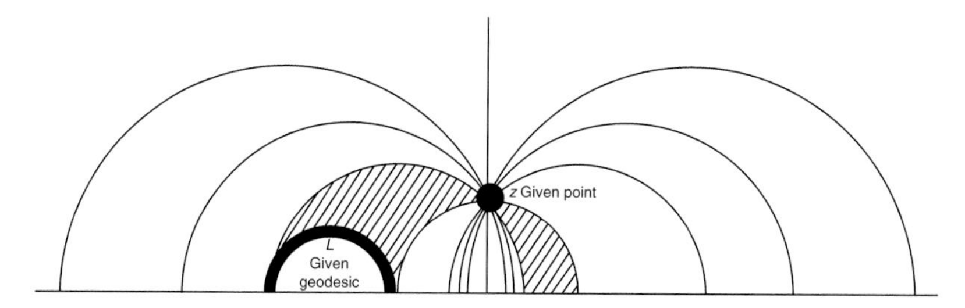

Figure 3.1 The failure of Euclid's fifth postulate is illustrated. All geodesics through z outside the hatched region fail to meet L (from [132], vol. I, p. 123).

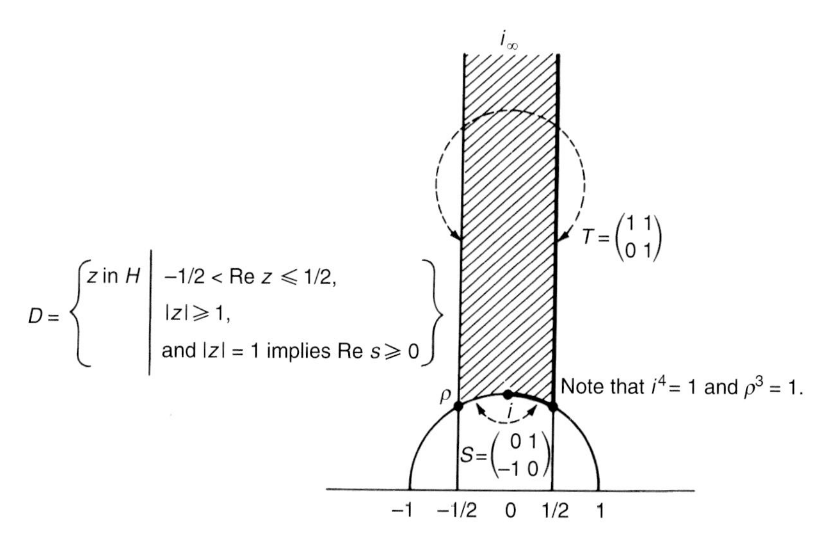

Figure 3.2 A non-Euclidean triangle D through the points $\rho,\ \rho+1, i_\infty$, which is a fundamental domain for H mod $\mathrm{SL}(2,\mathbb{Z})$. The domain D is hatched. The arrows show boundary identifications by fractional linear transformations from S and T, which generate $\mathrm{SL}(2,\mathbb{Z})/\{\pm I\}$ (from [132], p. 164).

Euclid's fifth postulate fails. There are infinitely many geodesics through a fixed point not meeting a given geodesic. See Figure 3.1.

The fundamental group Γ of M acts as a discrete group of distance-preserving transformations.

The favorite groups of number theorists are the **modular group** $\Gamma = \mathrm{SL}(2,\mathbb{Z})$ of 2×2 matrices having determinant 1 and integer entries and the quotient $\overline{\Gamma} = \Gamma/\{\pm I\}$.

We can identify the quotient $\mathrm{SL}(2,\mathbb{Z})\backslash H$ with the **fundamental domain** D pictured in Figure 3.2, where the sides are identified via $z \longrightarrow z+1$ and $z \longrightarrow -1/z$. These transformations generate $\overline{\Gamma}$.

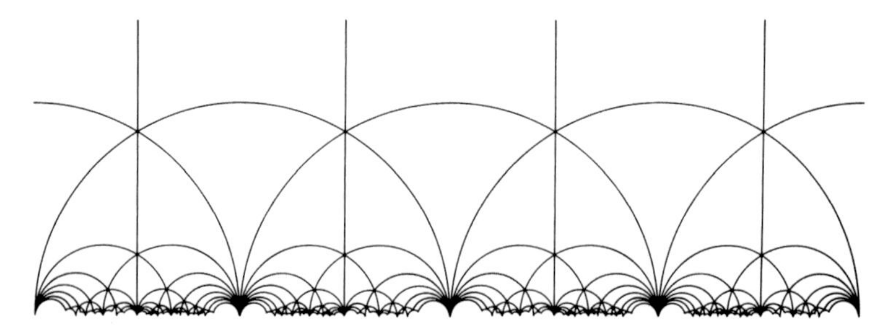

Figure 3.3 The tessellation of the upper half plane arising from applying the elements of the modular group to the fundamental domain D in the preceding figure. The figure is taken from [132], p. 166.

The images of D under elements of Γ provide a **tessellation** of the upper half plane. See Figure 3.3.

There are four types of elements of $\overline{\Gamma}$. They are determined by the Jordan form of the 2×2 matrix. The **corresponding fractional linear map will be one of four types**, as shown in the following table:

identity	$z \longrightarrow z$	
elliptic	$z \longrightarrow cz$	where $\|c\| = 1,\ c \neq 1$
hyperbolic	$z \longrightarrow cz$	where $c > 0,\ c \neq 1$
parabolic	$z \longrightarrow z + a$	

The **Riemann surface** $M = \Gamma\backslash H$ is compact and without **branch (ramification)** points if Γ has only the identity and hyperbolic elements. The modular group unfortunately has both elliptic and parabolic elements. It is easiest to deal with Selberg zeta functions when Γ has only the identity and hyperbolic elements. Examples of such groups are discussed in Svetlana Katok's book [67].

A hyperbolic element $\gamma \in \Gamma$ will have two fixed points on $\mathbb{R} \cup \{\infty\}$. Call these points z and w. Let $C(z, w)$ be a geodesic line or circle in H connecting points z and w in $\mathbb{R} \cup \{\infty\}$. Consider the image $\overline{C(z, w)}$ in the fundamental domain for $\Gamma\backslash H$. We say that $\overline{C(z, w)}$ is a **closed geodesic** if it is a closed curve in the fundamental domain (i.e., the beginning of the curve is at the same point as the end). A **primitive** closed geodesic is traversed only once. One can show that $\overline{C(z, w)}$ is a closed geodesic in $\Gamma\backslash H$ iff there is an element $\gamma \in \Gamma$ such that $\gamma C(z, w) \subset C(z, w)$. This means that z and w are the fixed points of

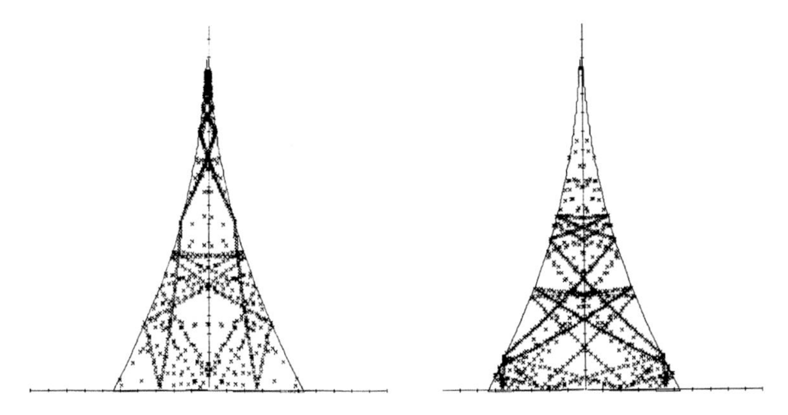

Figure 3.4 Images of points on two geodesic circles after they have been mapped into the fundamental domain of SL(2, $\mathbb{Z}$) and then mapped by Cayley transform into the unit disc. Here the geodesic circles have center 0 and radii $\sqrt{163}$ on the left, e on the right (from [132], vol. I, pp. 280–281).

a hyperbolic element of Γ. One can show that if a point q lies on $C(z, w)$ then so does γq, and the Poincaré distance between q and γq is $\log N\gamma$, where $N\gamma = a^2$, if γ has Jordan form $\begin{pmatrix} a & 0 \\ 0 & 1/a \end{pmatrix}$ with a real, $a \neq \pm 1$. Emil Artin was one of the first people to consider the question of whether these geodesics tend to fill up the fundamental domain as their length approaches infinity. This is related to ergodic theory and dynamical systems. It is also related to continued fractions. See Bedford *et al.* [13].

Exercise 3.1 Using a computer, graph $\overline{C(z, w)}$ for various choices of z, w. We did this in Figure 3.4 and then mapped everything into the unit disc using the **Cayley transform**

$$z \longrightarrow \frac{i(z-i)}{z+i}.$$

The lengths of these primitive closed geodesics coming from hyperbolic elements of Γ form the **length spectrum** of $M = \Gamma \backslash H$. These are our "primes" in M. Now we are ready to define the **Selberg zeta function** as

$$Z(s) = \prod_{[C]} \prod_{j \geq 1} \left(1 - e^{-(s+j)\nu(C)}\right). \tag{3.1}$$

The product is over all primitive closed geodesics C in $M = \Gamma\backslash H$ of length $\nu(C)$. Just as with the Ihara zeta function, this product can also be viewed as a product over conjugacy classes of primitive hyperbolic elements $\{\gamma\} \subset \Gamma$. Here by "primitive hyperbolic element" we mean that it generates its centralizer in Γ. The Selberg trace formula gives an explicit formula relating the spectrum of the Laplacian on M and the length spectrum of M.

The Selberg zeta function has many properties similar to the Riemann zeta function. Via the Selberg trace formula one can show that the logarithmic derivative of the Selberg zeta function has an analytic continuation as a meromorphic function. See Elstrodt [38], Patterson [98], and Bunke and Olbrich [22]. The non-trivial zeros of $Z(s)$ correspond to the discrete spectrum of the Poincaré Laplacian on $L^2(\Gamma\backslash H)$. This means that the Selberg zeta function satisfies the Riemann hypothesis (assuming that $\Gamma\backslash H$ is compact). Marie-France Vignéras [139] proved many properties of the Selberg zeta function for the non-compact quotient $\mathrm{SL}(2, \mathbb{Z})\backslash H$.

Sarnak has said that for non-arithmetic Γ with non-compact fundamental domain he doubts that one should think of $Z(s)$ as a zeta function, as he conjectures that the discrete spectrum of Δ is finite. Of course, this will be the case for the Ihara zeta function of a finite graph so perhaps some might think it is not a zeta at all.

To define "arithmetic" we must first define **commensurable subgroups** A, B of a group C; this means that $A \cap B$ has finite index both in A and B. Then suppose that Γ is an algebraic group over $\mathbb{Q}$, as in Borel's article in Borel and Mostow [20], p. 4. One says that Γ is **arithmetic** if there is a faithful rational representation ρ into the general linear group of $n \times n$ non-singular matrices such that ρ is defined over the rationals and $\rho(\Gamma)$ is commensurable with $\rho(\Gamma) \cap \mathrm{GL}(n, \mathbb{Z})$. Roughly we are saying that the integers are hiding somewhere in the definition of Γ. See [20] for more information. Arithmetic and non-arithmetic subgroups of $\mathrm{SL}(2, \mathbb{C})$ are discussed by Elstrodt, Grunewald, and Mennicke [39].

Exercise 3.2 Show that $Z(s+1)/Z(s)$ has a product formula which looks more like that for the Ihara zeta function than (3.1) does.

The theory of Ihara zeta functions for $(p^e + 1)$-regular graphs, p prime, can be understood via trace formulas for groups of 2×2 matrices over function fields; see Nagoshi [92]–[94]. One can also work out the theory of Ihara zetas on d-regular graphs based on trace formulas for groups acting on the d-regular tree. See Horton, Newland, and Terras [59], Terras [133], and Terras and Wallace [136].

4

Ruelle zeta function

Important references for the Ruelle zeta function include Ruelle [108] and Bedford, Keane, and Series [13]. Ruelle's motivation for his definition came partially from a paper by M. Artin and B. Mazur [4]. They were in turn inspired by the definition of the **zeta function of a projective non-singular algebraic variety** of dimension n defined over a finite field k with q elements. If N_m is the number of points of V with coordinates in the degree-m extension field of k, the zeta function of V is given by

$$Z(z, V) = \exp\left(\sum_{m=1}^{\infty} \frac{N_m}{m} z^m\right). \tag{4.1}$$

This zeta can be identified with that of a function field over k.

Example of varieties are given by taking solutions of polynomial equations over finite fields; e.g., $x^2 + y^2 = 1$ and $y^2 = x^3 + ax + b$. You actually have to look at the homogeneous version of the equations in projective space. For more information on these zeta functions, see Lorenzini [76], p. 280, or Rosen [104].

Note that N_m is the number of fixed points of F^m, where F denotes the **Frobenius morphism**, which takes a point with coordinates (x_i) to the point $\left(x_i^q\right)$. The **Weil conjectures**, ultimately proved in the general case by Deligne, say that

$$Z(z, V) = \prod_{j=0}^{2n} P_j(z)^{(-1)^{j+1}},$$

where the P_j are polynomials whose zeros have absolute value $q^{-j/2}$. Moreover the P_j have a cohomological meaning (roughly that $P_j(z) = \det(1 - zF^* | H^j(V))$. Here the Frobenius morphism has induced an action on the ℓ-adic étale cohomology. The case $n = 1$ is very similar to that of the Ihara zeta function for a $(q + 1)$-regular graph.

Artin and Mazur [4] replaced the Frobenius morphism of the algebraic variety with a diffeomorphism f of a smooth compact manifold M. They defined the set of fixed points of f^m to be

$$\mathrm{Fix}(f^m) = \left\{x \in M \middle| f^m(x) = x\right\}$$

and looked at the zeta function

$$\zeta(z) = \exp\left(\sum_{m=1}^{\infty} \frac{z^m}{m} \left| \mathrm{Fix}\left(f^m\right) \right|\right). \tag{4.2}$$

The Ruelle zeta function involves a function $f : M \to M$ on a compact manifold M. Assume that the set $\mathrm{Fix}(f^m)$ is finite for all $m \geq 1$. Suppose that $\varphi : M \to \mathbb{C}^{d \times d}$ is a matrix-valued function. The first type of **Ruelle zeta function** is defined by

$$\zeta(z) = \exp\left\{\sum_{m \geq 1} \frac{z^m}{m} \sum_{x \in \mathrm{Fix}(f^m)} \mathrm{Tr}\left(\prod_{k=0}^{m-1} \varphi\left(f^k(x)\right)\right)\right\}. \tag{4.3}$$

Here we consider only the special case for which $d = 1$ and φ is identically 1, when formula (4.3) looks exactly like formula (4.2). Ruelle also defines a second type of zeta function associated with a one-parameter semigroup of maps $f^t : M \to M$. See the reference above for the details.

Now we consider a special case in order to see that the Ihara zeta function of a graph is a Ruelle zeta function. For this we consider subshifts of finite type. Let I be a finite non-empty set (our alphabet). For a graph X, let I be the set of directed edges of X. Define the **transition matrix** $t = (t_{ij})_{ij \in I}$ to be a matrix of 0's and 1's.

For the case of a graph let t denote the **edge adjacency matrix** W_1 defined below.

Definition 4.1 For a graph X, define the **edge adjacency matrix** W_1 by orienting the m edges of X and labeling them as in formula (2.1). Then W_1 is the $2m \times 2m$ matrix with ij entry 1 if edge e_i feeds into e_j, provided that $e_j \neq e_i^{-1}$, and ij entry 0 otherwise. By "a **feeds into** b," we mean that the terminal vertex of edge a is the same as the initial vertex of edge b.

We will derive the **two-term determinant formula**

$$\zeta_X(u)^{-1} = \det(I - W_1 u)\,. \tag{4.4}$$

From this we shall later obtain the three-term determinant formula given in Theorem 2.5.

Note that the product $I^{\mathbb{Z}}$ is compact and thus so is the closed subset Λ defined by

$$\Lambda = \left\{ (\xi_k)_{k\in\mathbb{Z}} \middle| t_{\xi_k \xi_{k+1}} = 1, \text{ for all } k \right\}.$$

In the graph case, $\xi \in \Lambda$ corresponds to a path without backtracking.

A continuous function $\tau : \Lambda \to \Lambda$ such that $\tau(\xi)_k = \xi_{k+1}$ is called a **subshift of finite type**. In the graph case, this shifts the path left, assuming that the paths go from left to right.

Now we can find a new formula for the Ihara zeta function which shows that it is a Ruelle zeta. To understand this formula, we need a definition.

Definition 4.2 Let $N_m = N_m(X)$ be the **number of closed paths of length** m **without backtracking or tails in the graph** X.

From Definition 2.2, the definition of the Ihara zeta, we will prove in the next paragraph that

$$\log \zeta_X(u) = \sum_{m\geq 1} \frac{N_m}{m} u^m. \tag{4.5}$$

Compare this with formula (4.1) defining the zeta function of a projective variety over a finite field.

To prove formula (4.5), take the logarithm of Definition 2.2, where the product is over primes $[P]$ in the graph X, to obtain

$$\log \zeta_X(u) = \log \prod_{\substack{[P] \\ \text{prime}}} \left(1 - u^{\nu(P)}\right)^{-1} = -\sum_{[P]} \log\left(1 - u^{\nu(P)}\right)$$

$$= \sum_{[P]} \sum_{j\geq 1} \frac{1}{j} u^{j\nu(P)} = \sum_{P} \sum_{j\geq 1} \frac{1}{j\nu(P)} u^{j\nu(P)} = \sum_{P} \sum_{j\geq 1} \frac{1}{\nu(P^j)} u^{\nu(P^j)}$$

$$= \sum_{\substack{C \text{ closed} \\ \text{backtrackless} \\ \text{tailless path}}} \frac{1}{\nu(C)} u^{\nu(C)} = \sum_{m\geq 1} \frac{N_m}{m} u^m.$$

Here we have used the power series for $\log(1-x)$ to obtain the third equality. Then the fourth equality comes from the fact that there are $\nu(P)$ elements in the equivalence class $[P]$, for any prime $[P]$. The fifth equality comes from $\nu(P^j) = j\nu(P)$. The sixth equality is proved using the fact that any closed backtrackless tailless path C in the graph is a power of some prime path P. The last equality comes from Definition 4.2 of N_m.

If the subshift of finite type τ is as defined above for the graph X, we have

$$\left|\mathrm{Fix}(\tau^m)\right| = N_m. \tag{4.6}$$

It follows from this result and formula (4.5) that the Ihara zeta is a special case of the Ruelle zeta.

Next, using Definition 4.1, we claim that

$$N_m = \mathrm{Tr}\ W_1^m. \tag{4.7}$$

To see this, set $t = B = W_1$, with entries b_{ef} for oriented edges e, f. Then

$$\mathrm{Tr}\ W_1^m = \mathrm{Tr}\ B^m = \sum_{e_1,\ldots,e_m} b_{e_1e_2}b_{e_2e_3}\cdots b_{e_me_1},$$

where the sum is over all oriented edges of the graph. The b_{ef} are 0 unless edge e feeds into edge f without backtracking, i.e., the terminal vertex of e is the initial vertex of f and $f \neq e^{-1}$. Thus $b_{e_1e_2}b_{e_2e_3}\cdots b_{e_me_1} = 1$ means that the path $C = e_1e_2\cdots e_m$ is closed, backtrackless, tailless, and of length m. Formula (4.7) follows.

Then we use formulas (4.5) and (4.7) to see that

$$\log \zeta_X(u) = \sum_{m\geq 1} \frac{u^m}{m}\,\mathrm{Tr}\ W_1^m = \mathrm{Tr} \sum_{m\geq 1} \frac{u^m}{m} W_1^m$$

$$= \mathrm{Tr}\left(\log(I - uW_1)^{-1}\right) = \log\det(I - uW_1)^{-1}.$$

Here we have used the continuous linear property of the trace operator. Finally, we need the power series for the matrix logarithm and the following exercise.

Exercise 4.1 Show that $\exp \mathrm{Tr}\ A = \det \exp A$, for any matrix A. To prove this, you need to know that there is a non-singular matrix B such that $BAB^{-1} = T$ is upper triangular. See your favorite linear algebra book.

This proves formula (4.4) for the Ihara zeta function, which says that $\zeta_X(u) = \det(I - uW_1)^{-1}$. More generally, this is known as the Bowen–Lanford theorem for subshifts of finite type in the context of Ruelle zeta functions. The general result is as follows. The proof is similar to that for (4.4).

Proposition 4.3 (Bowen and Lanford) *One has the following formula for the Ruelle zeta function of a subshift of finite type τ with transition matrix t:*

$$\zeta(z) = \exp\left(\sum_{m\geq 1} \frac{z^m}{m}\left|\mathrm{Fix}(\tau^m)\right|\right) = \det(1 - zt)^{-1}.$$

Later, we will consider an edge zeta function attached to X which involves more than one complex variable and which has a similar determinant formula.

5
Chaos

References for the subject of chaos include Cipra [30], Haake [49], Miller and Takloo-Bighash [86] (see in particular the downloadable papers from the book's website at Princeton University Press), Rudnick [106], [107], Sarnak [110], Terras [134] and [135].

Quantum chaos is in part the study of the statistics of energy levels of quantum mechanical systems, i.e., the eigenvalues of the Schrödinger operator $\mathcal{H}$, where $\mathcal{H}\phi = E\phi$. A good website for quantum chaos is that of Matthew W. Watkins:

www.maths.ex.ac.uk/~mwatkins.

We quote Oriol Bohigas and Marie-Joya Gionnoni [16], p. 14: "The question now is to discover the stochastic laws governing sequences having very different origins, as illustrated ... There are displayed six spectra, each containing 50 levels ..." (see Figure 5.1). Note that in Figure 5.1 the spectra have been rescaled to the same vertical axis from 0 to 49 and that we have added two more columns to the original figure.

In Figure 5.1, column (a) represents a Poisson spectrum, meaning that of a random variable with spacings of probability density e^{-x}. Column (b) represents primes between 7 791 097 and 7 791 877. Column (c) represents the resonance energies of the compound nucleus observed in the nuclear reaction n + ^{166}Er. Column (d) comes from eigenvalues corresponding to the transverse vibrations of a membrane whose boundary is the Sinai billiard, which is a square with a circular hole cut out centered at the center of the square. Column (e) is from the positive imaginary part of zeros of the Riemann zeta function from the 1551th to the 1600th zero. Column (f) is equally spaced – the picket-fence or uniform distribution. Column (g) comes from Sarnak [110] and corresponds to eigenvalues of the Poincaré Laplacian on the fundamental

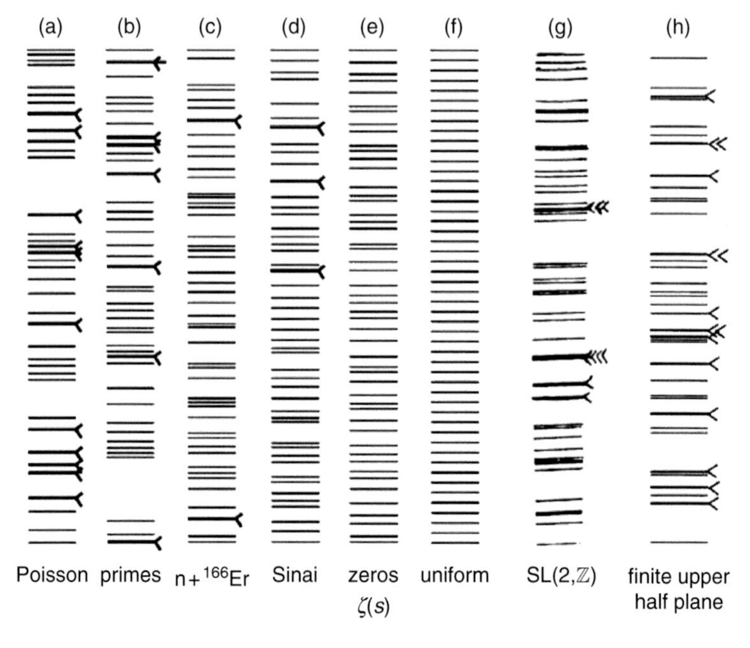

Figure 5.1 The level spacings in columns (a)–(f) are from Bohigas and Giannoni [16] and column (g) is from Sarnak [110]. Each column is a spectrum segment containing 50 levels. The arrowheads mark the occurrences of pairs of levels with spacing smaller than 1/4. The labels are explained in the text. Column (h) contains the finite upper half plane graph eigenvalues (without multiplicity) for the prime 53, with $\delta = a = 2$.

domain of the modular group SL(2, $\mathbb{Z}$) consisting of 2×2 integer matrices of determinant 1. From the point of view of randomness, columns (g) and (h) should be moved to lie next to column (b). Column (h) is the spectrum of a finite upper half plane graph for $p = 53$ ($a = \delta = 2$), without multiplicity. See Terras [133] for the definition of finite upper half plane graphs.

Exercise 5.1 Produce your own versions of as many columns of Figure 5.1 as possible for poles or zeros of various zeta functions or eigenvalues of various matrices or operators.

Quantum mechanics says that the energy levels E of a physical system are the eigenvalues of a Schrödinger equation $\mathcal{H}\phi = E\phi$, where $\mathcal{H}$ is the Hamiltonian (a differential operator), ϕ is the state function (an eigenfunction of $\mathcal{H}$), and E is a corresponding energy level (an eigenvalue of $\mathcal{H}$). For complicated systems, it is usually impossible to know all the energy levels. So instead physicists investigate the statistical theory of these energy levels.

This approach is used in ordinary statistical mechanics as well. Of course symmetry groups (i.e., groups of operators commuting with $\mathcal{H}$) have a big effect on the energy levels.

In the 1950s Wigner (see [143]) considered modelling $\mathcal{H}$ with large real symmetric $n \times n$ matrices whose entries are independent Gaussian random variables. He found that the histogram of the eigenvalues looks like a semicircle (or, more precisely, a semi-ellipse). It has been named the **Wigner semicircle distribution**. For example, he considered the eigenvalues of 197 "random" real symmetric 20×20 matrices. The upper histogram in Figure 5.2 shows the results of an analog of Wigner's experiment using Matlab.

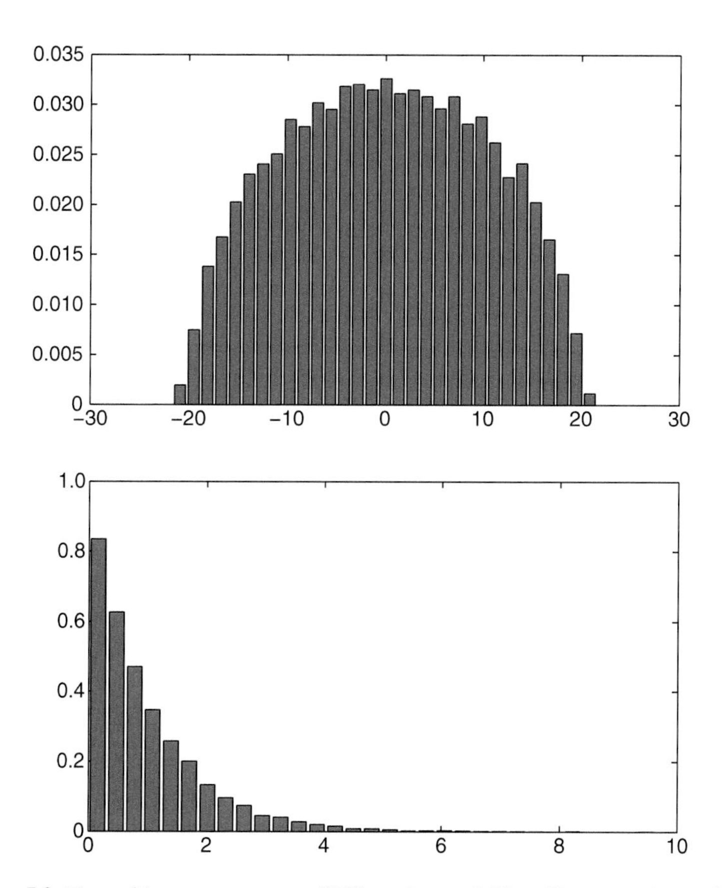

Figure 5.2 *Upper histogram*, spectra of 200 random real 50×50 symmetric matrices created using Matlab. *Lower histogram*, spacings of the spectra of the matrices from the top histogram. The distribution appears to be e^{-x}, the same as that for the spacings of Poisson random variables.

We took 200 random (normally distributed) real symmetric 50×50 matrices with entries chosen according to the normal distribution. Wigner notes on p. 5 of [143]: "What is distressing about this distribution [of the eigenvalues] is that it shows no similarity to the observed distribution in spectra." This may be the case in physics, but the semicircle distribution is well known to number theorists as the Sato–Tate distribution.

Exercise 5.2 Repeat the experiment that produced Figure 5.2 using uniformly distributed matrices rather than normally distributed matrices. This is a problem best done with Matlab which has the commands `rand(50)`, giving a random 50×50 matrix with uniformly distributed entries, and `randn(50)`, giving a normal random 50×50 matrix. You must normalize the eigenvalues to have mean spacing 1 by considering them in batches.

So, physicists have devoted more attention to histograms of level spacings rather than levels. This means that you arrange the energy levels (eigenvalues) E_i in decreasing order:

$$E_1 \geq E_2 \geq \cdots \geq E_n.$$

Assume that the eigenvalues are normalized in such a way that the mean of the level spacings $|E_i - E_{i+1}|$ is 1. Then one can ask for the shape of the histogram of the normalized level spacings. There are (see Sarnak [110]) two main sorts of answer to this question: the distribution for **Poisson-level spacings**, e^{-x}, and that for **Gaussian orthogonal ensemble (GOE) spacings** (see Mehta [85]) which is more complicated to describe exactly but looks like $\frac{1}{2}\pi x e^{-\pi x^2/4}$ (the **Wigner surmise**). In 1957 Wigner (see [143]) gave an argument for the surmise that the level spacing histogram for levels having the same values of all quantum numbers is given by $\frac{1}{2}\pi x e^{-\pi x^2/4}$ if the mean spacing is 1. In 1960 Gaudin and Mehta found the correct distribution function, which is surprisingly close to Wigner's conjecture but different. The correct curves are labeled GOE (Gaussian orthogonal ensemble) in Figure 5.4. Note the level repulsion indicated by the vanishing of the function at the origin. Also in Figure 5.4 we see the Poisson spacing density e^{-x}. See Bulmer [21], p. 102.

The spacing histogram in Figure 5.2 is seen in the lower half of the figure and it looks Poisson. It should be compared with the lower histogram in Figure 5.3, which represents the level spacings of one 1001×1001 matrix. This illustrates an important aspect of the dichotomy between GOE and Poisson behavior. If you throw lots of random symmetric matrices together you get Poisson spacing but if you take just one large symmetric matrix you see GOE spacing. Later we will say more about this dichotomy.

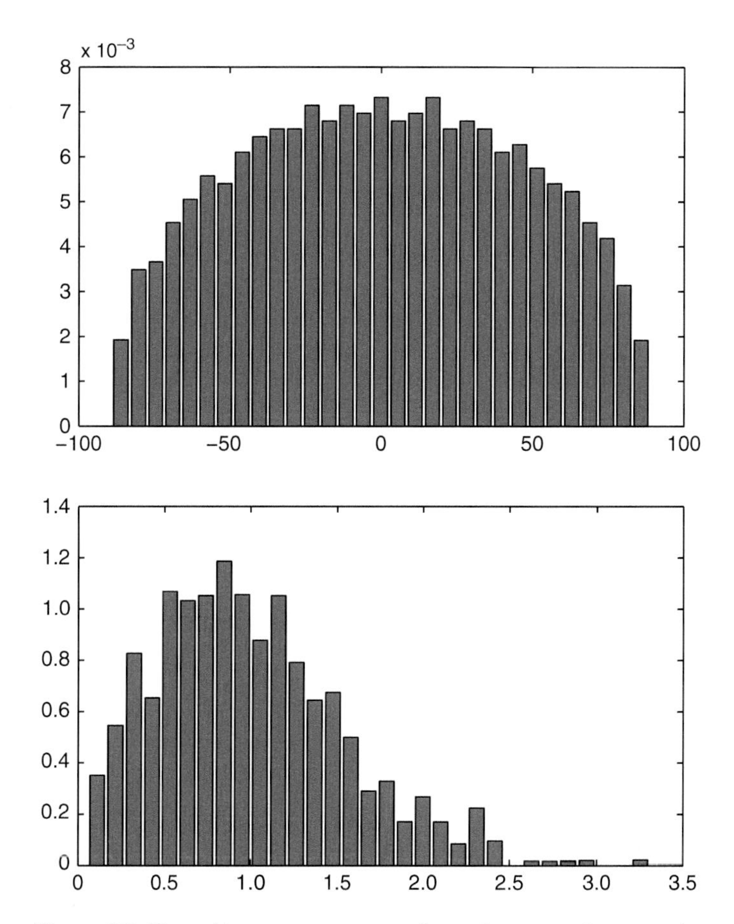

Figure 5.3 *Upper histogram*, spectrum of a random normal symmetric real 1001×1001 matrix. *Lower histogram*, normalized level spacings for the same spectrum.

Sarnak [110], p. 160, states:

> It is now believed that for integrable systems the eigenvalues follow the Poisson behavior while for chaotic systems they follow the GOE distribution.

The GOE is formed by the eigenvalues of a random $n \times n$ symmetric real matrix as n goes to infinity, whereas the GUE is formed by the eigenvalues of a random complex $n \times n$ Hermitian matrix as n goes to infinity.

There have been many experimental studies comparing GOE predictions and nuclear data. Work on the atomic spectra and spectra of molecules also exists. In Figure 5.4, we reprint a figure of Bohigas, Haq, and Pandey [17] giving a

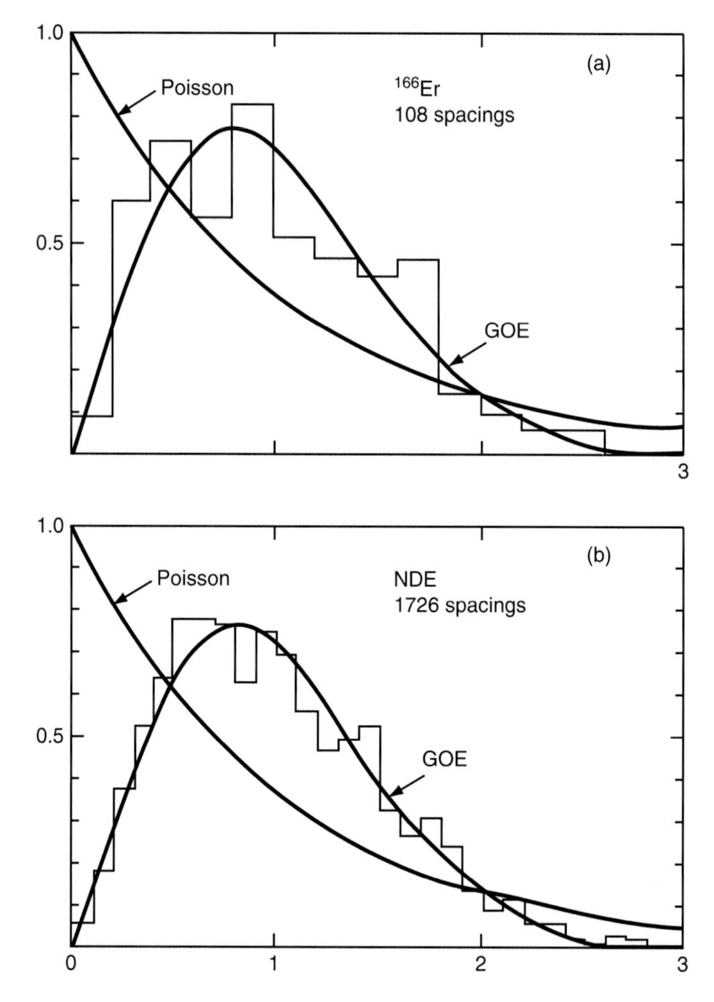

Figure 5.4 (Bohigas, Haq, and Pandey [17]) Level spacing histograms for (a) ^{166}Er and (b) a nuclear data ensemble.

comparison of histograms of level spacings for ^{166}Er (upper) and a nuclear data ensemble (NDE) consisting of about 1700 energy levels corresponding to 36 sequences of 32 different nuclei (lower). Bohigas *et al.* state: "The criterion for inclusion in the NDE is that the individual sequences be in general agreement with GOE."

Andrew Odlyzko (see www.dtc.umn.edu/~odlyzko/doc/zeta.html) has investigated the level spacing distribution for the non-trivial zeros γ_n of the Riemann zeta function. He considers only zeros which are high up on the

Re $s = 1/2$ line. Assume the Riemann hypothesis and look at the zeros ordered by imaginary part, i.e.,

$$\{\gamma_n | \zeta\,(1/2 + i\gamma_n) = 0,\, \gamma_n > 0\}.$$

For the normalized level spacings, replace γ_n by $\widetilde{\gamma_n} = (1/2\pi)\gamma_n \log \gamma_n$, since we want the mean spacing to be 1. Here one needs to know that the number of γ_n such that $\gamma_n \leq T$ is asymptotic to $(1/2\pi)T \log T$ as $T \longrightarrow \infty$.

Historically, the connections between the statistics of the Riemann zeta zeros γ_n and the statistics of the energy levels of quantum systems were made in a dialogue between Freeman Dyson and Hugh Montgomery over tea at the Institute for Advanced Study, Princeton. Odlyzko's experimental results show that the level spacings $|\gamma_n - \gamma_{n+1}|$, for large n, look like that of the **Gaussian unitary ensemble** (**GUE**), i.e., the eigenvalue distribution of a random complex Hermitian matrix. See Figure 5.5.

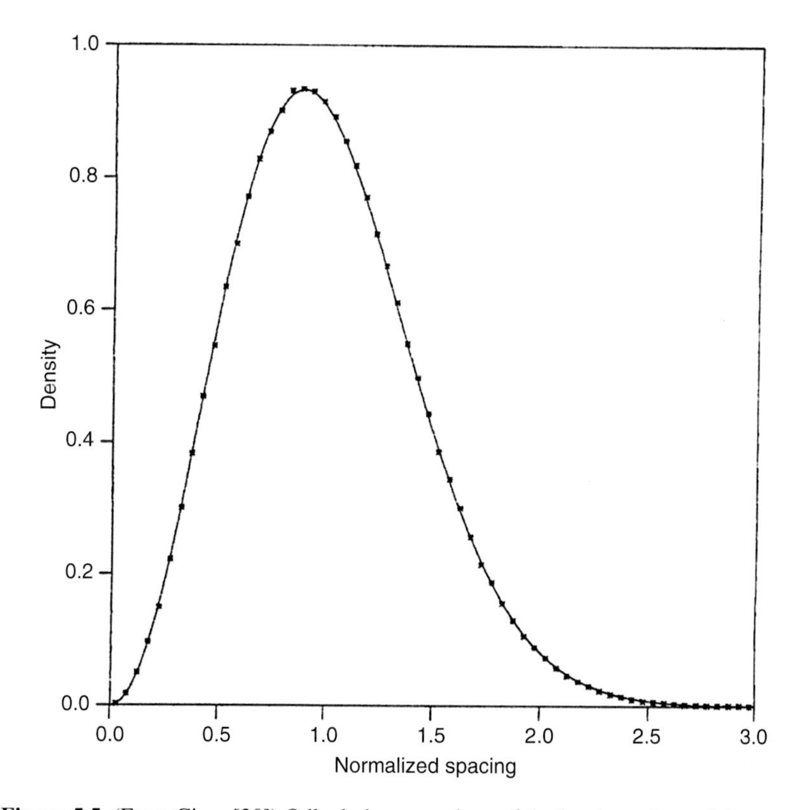

Figure 5.5 (From Cipra [30]) Odlyzko's comparison of the level spacings of the zeros of the Riemann zeta function (points) and the GUE spacings (curve). See Forrester and Odlyzko [41]. The fit is good for the 1 041 000 zeros near the 2×10^{20} zero.

The level spacing distribution for the eigenvalues of Gaussian unitary matrices (the GUE) is not a standard function in Matlab, Maple, or Mathematica. Sarnak [110] and Katz and Sarnak [69] proceeded as follows. Let $K_s : L^2[0, 1] \longrightarrow L^2[0, 1]$ be the integral operator with kernel defined by

$$h_s(x, y) = \frac{\sin \frac{1}{2}\pi s(x - y)}{\frac{1}{2}\pi(x - y)}, \qquad \text{for } s \geq 0.$$

Approximations to this kernel have been investigated in connection with the uncertainty principle (see Terras [132], vol. I, p. 51). The eigenfunctions are spheroidal wave functions. Let $E(s)$ be the Fredholm determinant $\det(I - K_s)$ and let $p(s) = E''(s)$. Then $p(s) \geq 0$ and $\int_0^\infty p(s)ds = 1$. The Gaudin–Mehta distribution ν is defined in the GUE case by $\nu(I) = \int_I p(x)ds$. For the GOE case the kernel h_s is replaced by $\{h_s(x, y) + h_s(-x, y)\}/2$. See also Mehta [85]. You can find a Mathematica program to compute the GUE function at the website in [41].

Katz and Sarnak and others (see [69], p. 23) have investigated many zeta and L-functions of number theory and have found that "the distribution of the high zeroes of any L-function follows the universal GUE laws, while the distribution of the low-lying zeroes of certain families follow the laws dictated by symmetries associated with the family. The function field analogs of these phenomena can be established ..." More precisely (see [69], p. 11) they show that "the zeta functions of almost all curves C $\left[\text{over a finite field } \mathbb{F}_q\right]$ satisfy the Montgomery–Odlyzko law [the GUE] as q and g [the genus] go to infinity."

These statistical phenomena are as yet unproved for most zeta functions of number theory, e.g., Riemann's. But the experimental evidence of Rubinstein [105] and Strömbergsson [126] and others is strong. Figures 3 and 4 of Katz and Sarnak [69] show the level spacings for the zeros of the L-function corresponding to the modular form Δ and for the zeros of the L-function corresponding to a certain elliptic curve, for comparison with the GUE. Strömbergsson's web site has similar pictures for L-functions corresponding to Maass wave forms (http://www.math.uu.se/~andreas/zeros.html). All these pictures look like the GUE. See also the L-function wiki.

The conjectured dichotomy of quantum chaos is illustrated in Table 5.1. The first column gives those spacings conjectured to be of random matrix theory (RMT) type and the second column gives those spacings conjectured to be Poisson. In first row we have the Bohigas–Giannoni–Schmit conjecture from 1984 on the left and the Berry–Tabor–Gutzwiller conjecture from 1977 on the right. Sarnak invented the term "arithmetical quantum chaos" to describe the second row of the table. Recall the definition of the arithmetic group from Chapter 3. Using terminology from André Weil's Columbia University lectures

Table 5.1 *Conjectural dichotomy of quantum chaos*

RMT spacings (GOE etc.)	Poisson spacings
Quantum spectra of a system with chaos in classical counterpart	Energy levels of a quantum system with an integrable system for classical counterpart
Eigenvalues of the Laplacian for a non-arithmetic manifold	Eigenvalues of the Laplacian for an arithmetic manifold
Zeros Riemann zeta	
Poles of the Ihara zeta of a random regular graph	Poles of the Ihara zeta of a Cayley graph of an abelian group

in 1971, the Laplacian on a Riemann surface $\Gamma \backslash H$ can "smell" number theory, namely $\mathbb{Z}$, if it is present somewhere in Γ.

Figure 5.3 showed the Wigner semicircle distribution for the spectrum of a large random symmetric real matrix. When the matrix is the adjacency matrix of a large regular graph under suitable hypotheses, surprisingly one also sees an approximate semicircle. This was proved by McKay [84]. See Theorem 5.3 below. In [59] Horton, Newland, and Terras gave a proof due to Nagoshi [92], [94] which uses the Selberg trace formula on the $(q+1)$-regular tree and the Weyl equidistribution theorem. See also Fan Chung *et al.* [28], Mehta [85], and Sunada [130].

Before stating McKay's result, we need to recall Weyl's equidistribution criterion (see Weyl [142] or Iwaniec and Kowalski [64], Chapter 21).

Definition 5.1 A sequence $\{x_n\}$ in an interval I on the real line is said to be **equidistributed** with respect to a measure $d\mu$ iff, for every open set B in interval I,

$$\lim_{N\to\infty} \frac{1}{N} \left| \{ n \leq N | x_n \in B \} \right| = \mu(B).$$

This equidistribution property is equivalent to the statement that, for any continuous function with compact support f on I,

$$\lim_{N\to\infty} \frac{1}{N} \sum_{n\leq N} f(x_n) = \int_I f(x) d\mu(x). \tag{5.1}$$

Definition 5.2 The **Plancherel measure** $d\mu_q$ on $[-2\sqrt{q}, 2\sqrt{q}\,]$ is given by

$$d\mu_q = \frac{q+1}{2\pi} \frac{\sqrt{4q-\lambda^2}}{(q+1)^2-\lambda^2} d\lambda.$$

The measure $d\mu_q$ is that of the Plancherel theorem for the $(q+1)$-regular tree as in [59], where a proof of the following theorem is to be found.

Theorem 5.3 (McKay [84]) *Let $\{X_m\}_{m\geq 1}$ be a sequence of $(q+1)$-regular graphs such that for each $r>0$ we have*

$$\lim_{n\to\infty} \frac{N_r(X_n)}{|V(X_n)|} = 0, \qquad \text{where } N_r(X_n) \text{ is from Definition 4.2.}$$

Then, if A_{X_m} is the adjacency matrix of X_m, the spectrum of A_{X_m} becomes equidistributed with respect to the measure $d\mu_q$ on $[-2\sqrt{q}, 2\sqrt{q}\,]$, from Definition 5.2, as $n\to\infty$. More explicitly, this means that if $[\alpha,\beta]\subset[-2\sqrt{q}, 2\sqrt{q}\,]$ then

$$\lim_{m\to\infty} \frac{\#\left\{\lambda \in \mathrm{Spec}\left(A_{X_m}\right) | \alpha \leq \lambda \leq \beta\right\}}{|X_m|} = \frac{q+1}{2\pi}\int_\alpha^\beta \frac{\sqrt{4q-\lambda^2}}{(q+1)^2-\lambda^2}\, d\lambda.$$

Derek Newland [95] investigated the spacings of the poles of the Ihara zeta function for various kinds of graph. He found that the pole spacings for large random k-regular graphs appear to be derived from the adjacency matrix eigenvalue spacings, being GOE. However, the pole spacings for the Cayley graphs of abelian groups (in particular, the Euclidean graphs in Section 9.2 and in Terras [133]) appear to be spacings of Poisson random variables. Newland's experiments give the last row of Table 5.1, the conjectural dichotomy table, namely the spacings of poles of the Ihara zeta of a random regular graph on the left (which are essentially GOE) and the spacings of the poles of the Ihara zeta of a Cayley graph of an abelian group on the right (Poisson). Figure 5.6 shows the result of one such Mathematica experiment for a random regular graph. See Skiena [114] for more information on the way Mathematica deals with graphs. See also the article of Jakobson, Miller, Rivin, and Rudnick in [53], pp. 317–327.

Note For the level spacing to look GOE or GUE one expects the zeta function not to be a product of other zetas or L-functions. The zeta function of a Cayley graph such as the Euclidean graph shown in the lower part of Figure 5.6 will be a product over the representations of the group. Thus it behaves more like a product of different, unrelated, zeta functions rather than one zeta. This is the sort of behavior we saw when comparing Figures 5.2 and 5.3. See Farmer [40] for some information about random matrix theory and families of L-functions.

Exercise 5.3 Compute the change of variables between an element of the spectrum of the adjacency matrix and the imaginary part of s when q^{-s} is a pole of ζ_X for a $(q+1)$-regular graph X.

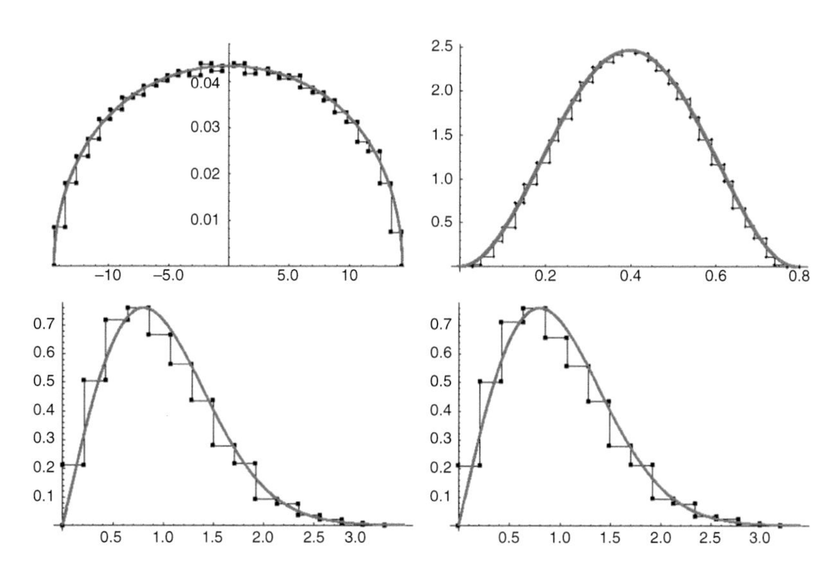

Figure 5.6 (Taken from Newland [95]) For a pseudo-random regular graph with degree 53 and 2000 vertices, generated by Mathematica, the upper row shows the distributions for the eigenvalues of the adjacency matrix (*left*) and the imaginary parts of the Ihara zeta poles (*right*). The lower row contains their respective level spacings; the smooth curve on the left is the Wigner surmise for the GOE $y = (\pi x/2)e^{-\pi x^2/4}$.

Exercise 5.4 Do an experiment similar to that of Figure 5.6 for a Cayley graph of your choice or for an n-cover of $X = K_4 - e$, where e is an edge.

As a final project for Part I, you might try to list all the zeta functions that you can find and figure out what they are good for. There are lots of them. We have left out zeta functions of codes, for example.

Later we will consider the spacings of poles of zetas for irregular graphs and zetas of covering graphs.

Part II

Ihara zeta function and the graph theory prime number theorem

Graph theory zetas first appeared in work of Ihara on p-adic groups in the 1960s (see [62]). Serre [113] made the connection with graph theory. The main authors on the subject in the 1980s and 1990s were Sunada [128–130], Hashimoto [50], [51], and Bass [12]. Other references are Venkov and Nikitin [138] and Northshield's paper in the volume of Hejhal *et al.* [53]. The main properties of the Riemann zeta function have graph theory analogs, at least for regular graphs. For irregular graphs there is no known functional equation and it is difficult to formulate the Riemann hypothesis, but we will try.

Much of our discussion can be found in the papers of the author and Harold Stark [119], [120], [122]. Part II will include the story of Ihara zetas of regular graphs, the connection between the Riemann hypothesis and expanders, and the graph theory prime number theorem.

We do not consider zeta functions of infinite graphs here; such zeta functions are discussed by, for example, Clair and Mokhtari-Sharghi [31], Grigorchuk and Zuk [46], and Guido, Isola, and Lapidus [48]. Nor do we consider directed graphs; zeta functions for such graphs are discussed by, for example, Horton [57], [58]. There are also extensions to hypergraphs (see Storm [124]) and buildings (see Ming-Hsuan Kang, Wen-Ching Winnie Li, and Chian-Jen Wang [66]).

Throughout Part II we will assume Theorem 2.5, Ihara's theorem. It will be proved in Part III, where we will also consider the multivariable zeta functions known as edge and path zeta functions of graphs. We will show how to specialize the path zeta to the edge zeta and the edge zeta to the original one-variable Ihara (vertex) zeta.

6
Ihara zeta function of a weighted graph

Many applications involve weighted or metric graphs, that is, graphs with positive real numbers attached to the edges to represent length or resistance or some other physical attribute. In particular, quantum graphs are weighted (see [14]). Other references for weighted graphs are Chung and Yau [29] or Osborne and Severini [96]. For the most part we will not consider weighted graphs here but let us at least give a natural extension of the definition of the Ihara zeta function to weighted graphs.

Definition 6.1 For a graph X with oriented edge set $\overrightarrow{E}$, consisting of $2|E|$ oriented edges, suppose that we have a weighting function $L : \overrightarrow{E} \to \mathbb{R}^+$. Then define the **weighted length** of a closed path $C = a_1 a_2 \cdots a_s$, where $a_j \in \overrightarrow{E}$, by

$$\nu(C, L) = \nu_X(C, L) = \sum_{i=1}^{s} L(a_i).$$

Definition 6.2 The **Ihara zeta function of a weighted (undirected) graph** for $|u|$ small and $u \notin (-\infty, 0)$ is

$$\zeta_X(u, L) = \prod_{[P]} \left(1 - u^{\nu(P,L)}\right)^{-1}.$$

Clearly, when $L = 1$, meaning the function such that $L(e) = 1$ for all edges e in X, we have $\zeta_X(u, 1) = \zeta_X(u)$, our original Ihara zeta function.

Definition 6.3 Given a graph X with positive integer-valued weight function L, define the **inflated graph** X_L in which each edge e is replaced by an edge with $L(e) - 1$ new degree-2 vertices.

Then clearly $\nu_X(C, L) = \nu_{X_L}(C, 1)$, where the argument 1 means again that $L(e) = 1$ for all edges e. It follows that **for positive integer-valued weights** L

we have the following identity relating the weighted zeta and the ordinary Ihara zeta:

$$\zeta_X(u, L) = \zeta_{X_L}(u).$$

Therefore $\zeta_X(u, L)^{-1}$ is a polynomial for integer-valued weights L.

For non-integer weights, it is possible to obtain a determinant formula using the edge zeta functions in Chapter 11. See also Horton *et al.* [61] or Mizuno and Sato [89].

Example 6.4: Inflation of K_5 Suppose that $Y = K_5$, the complete graph on five vertices. Let $L(e) = 5$ for each of the 10 edges of X. Then $X = Y_L$ is the graph on the left in Figure 8.1. The new graph X has 45 vertices (four new vertices on the 10 edges of K_5). One sees easily that

$$\zeta_X(u)^{-1} = \zeta_{K_5}\left(u^5\right)^{-1} = \left(1 - u^{10}\right)^5\left(1 - 3u^5\right)\left(1 - u^5\right)\left(1 + u^5 + 3u^{10}\right).$$

Exercise 6.1 What happens to $\zeta_Y(u)$ if $Y = X_{L_n}$ for $L_n = n$ as $n \to \infty$?

For the most part, we shall restrict our discussion to non-weighted graphs from now on.

7

Regular graphs, location of poles of the Ihara zeta, functional equations

Next we want to consider the Ihara zeta function for regular graphs (which are unweighted and which satisfy our usual hypotheses, for the most part). We need some facts from graph theory first. References for the subject include Biggs [15], Bollobas [18], [19], Cvetković, Doob, and Sachs [32].

Definition 7.1 A graph is a **bipartite graph** iff its set of vertices can be partitioned into two disjoint sets S, T such that no vertex in S is adjacent to any other vertex in S and no vertex in T is adjacent to any other vertex in T.

Exercise 7.1 Show that the cube of Figure 2.8 is an example of a bipartite graph.

Proposition 7.2 (Facts about Spectrum A, when A is the adjacency operator of a connected $(q+1)$-regular graph X) *Assume that X is a connected $(q+1)$-regular graph and that A is its adjacency matrix. Then:*

(1) $\lambda \in \text{Spectrum } A$ *implies that* $|\lambda| \leq q+1$*;*
(2) $q+1 \in \text{Spectrum } A$ *and has multiplicity 1;*
(3) $-(q+1) \in \text{Spectrum } A$ *iff the graph X is bipartite.*

Proof of fact (1) Note that $q+1$ is clearly an eigenvalue of A corresponding to the constant vector. Suppose that $Av = \lambda v$, for some column vector $v = {}^t(v_1 \cdots v_n) \in \mathbb{R}^n$, and suppose that the maximum $|v_i|$ occurs at $i = a$. Then, using the notation $b \sim a$ to mean that the bth vertex is adjacent to the ath, we have

$$|\lambda|\,|v_a| = \left|(Av)_a\right| = \left|\sum_{b \sim a} v_b\right| \leq (q+1)|v_a|.$$

Fact (1) follows.

Proof of fact (2) Suppose that $Av = (q+1)v$ for some non-zero vector $v = {}^{\mathrm{t}}(v_1, \ldots, v_n) \in \mathbb{R}^n$. Again suppose that the maximum $|v_i|$ occurs at $i = a$. We can assume $v_a > 0$, since we can multiply the vector v by -1 if necessary. As in the proof of fact (1),

$$\begin{aligned}(q+1)v_a &= (Av)_a \\ &= \sum_{b \sim a} v_b \le (q+1)v_a.\end{aligned}$$

To have equality, there can be no cancellation in this sum and therefore $v_b = v_a$ for each b adjacent to a. Since we have assumed that X is connected, we can iterate this argument and conclude that v must be the constant vector.

Exercise 7.2 (a) Prove fact (3) above.
(b) Show that if $q+1$ has multiplicity 1 as an eigenvalue of the adjacency matrix of a $(q+1)$-regular graph then this graph must be connected.

Definition 7.3 Suppose that X is a connected $(q+1)$-regular graph (without degree-1 vertices). We say that the Ihara zeta function $\zeta_X(q^{-s})$ satisfies the **Riemann hypothesis** iff, when $0 < \operatorname{Re} s < 1$,

$$\zeta_X(q^{-s})^{-1} = 0 \qquad \Longrightarrow \qquad \operatorname{Re} s = 1/2.$$

Note that, if $u = q^{-s}$, $\operatorname{Re} s = \frac{1}{2}$ corresponds to $|u| = 1/\sqrt{q}$.

Theorem 7.4 *For a connected $(q+1)$-regular graph X, $\zeta_X(u)$ satisfies the Riemann hypothesis iff the graph X is Ramanujan in the sense of Definition 2.6.*

Proof Use Theorem 2.5 to see that

$$\zeta_X\left(q^{-s}\right)^{-1} = \left(1-u^2\right)^{r-1} \prod_{\lambda \in \text{Spectrum } A} \left(1 - \lambda u + qu^2\right).$$

Write $1 - \lambda u + qu^2 = (1-\alpha u)(1-\beta u)$, where $\alpha\beta = q$ and $\alpha + \beta = \lambda$. Note that α, β are the reciprocals of poles of $\zeta_X(u)$. Using the facts in Proposition 7.2 above, we have three cases.

Case 1 $\lambda = \pm(q+1)$, implying that $\alpha = \pm q$ and $\beta = \pm 1$.
Case 2 $|\lambda| \le 2\sqrt{q}$, implying that $|\alpha| = |\beta| = \sqrt{q}$.
Case 3 $2\sqrt{q} < |\lambda| < q+1$, implying that $\alpha, \beta \in \mathbb{R}$ and $1 < |\alpha| = |\beta| < q$, $|\alpha| = |\beta| \ne \sqrt{q}$.

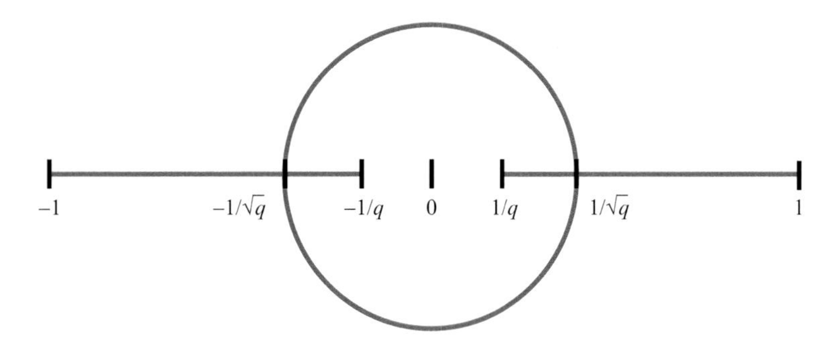

Figure 7.1 The possible locations for the poles of $\zeta_X(u)$ for a regular graph are the circle and the real-axis sections shown. The circle corresponds to the part of the spectrum of the adjacency matrix satisfying the Ramanujan inequality. The real poles correspond to the non-Ramanujan eigenvalues of A, except for the poles on the circle itself and those at the endpoints of the intervals.

To deduce these results, let u be either α^{-1} or β^{-1}. Then, by the quadratic formula, we have α or $\beta = u^{-1}$ where

$$u = \left(\lambda \pm \sqrt{\lambda^2 - 4q}\right)/2q.$$

Cases 1 and 2 are easily seen. We leave them as **exercises**. To understand case 3, first assume that $\lambda > 0$ and note that $u = \left(\lambda + \sqrt{\lambda^2 - 4q}\right)/(2q)$ is a monotone increasing function of λ. This implies that the larger root u is in the interval $\left(1/\sqrt{q}, 1\right)$. Where does the smaller root $u' = \left(\lambda - \sqrt{\lambda^2 - 4q}\right)/(2q)$ lie? Answer: $\left|u'\right| \in \left(1/q, 1/\sqrt{q}\right)$. Here we have used the fact that $uu' = 1/q$. A similar argument works for negative λ (**exercise**). The proof of the theorem is concluded by noting that when $u = q^{-s}$, case 2 gives $\operatorname{Re} s = 1/2$. □

Figure 7.1 shows the possible locations of the poles of the Ihara zeta function of a $(q+1)$-regular graph. The poles satisfying the Riemann hypothesis are those on the circle; the circle basically corresponds to case 2 in the preceding proof. The real axis sections correspond to cases 1 and 3.

Exercise 7.3 Fill in the details in the proof of Theorem 7.4. Then show that Figure 7.1 gives the possible locations of the poles of the Ihara zeta function of a $(q+1)$-regular graph. Label the locations corresponding to the three cases in the proof.

Exercise 7.4 Show that the radius of convergence of the product defining the Ihara zeta function of a $(q+1)$-regular graph is $R_X = 1/q$.

The following proposition gives some functional equations for the Ihara zeta function corresponding to a regular graph. If we set $u = q^{-s}$, the functional equations relate the value at s to that at $1-s$, just as in the case of the Riemann zeta function.

Proposition 7.5 *Suppose that X is a $(q+1)$-regular connected graph without degree-1 vertices and with $n=|V|$. Then we have the following* functional equations, *among others.*

(1) $\Lambda_X(u) \equiv (1-u^2)^{r-1+n/2}(1-q^2u^2)^{n/2}\,\zeta_X(u) = (-1)^n\Lambda_X(1/qu)$.
(2) $\xi_X(u) \equiv (1+u)^{r-1}\,(1-u)^{r-1+n}\,(1-qu)^n\,\zeta_X(u) = \xi_X(1/qu)$.
(3) $\Xi_X(u) \equiv (1-u^2)^{r-1}\,(1+qu)^n\,\zeta_X(u) = \Xi_X\big(1/qu\big)$.

Proof We will prove part (1) and leave the rest as an **exercise**. To prove part (1), write

$$\begin{aligned}\Lambda_X(u) &= \left(1-u^2\right)^{n/2}\left(1-q^2u^2\right)^{n/2}\det\left(I-Au+qu^2I\right)^{-1}\\ &= \left(\frac{q^2}{q^2u^2}-1\right)^{n/2}\left(\frac{1}{q^2u^2}-1\right)^{n/2}\det\left(I-A\frac{1}{qu}+\frac{q}{(qu)^2}I\right)^{-1}\\ &= (-1)^n\Lambda_X\left(\frac{1}{qu}\right).\end{aligned}$$

□

Exercise 7.5 Prove parts (2) and (3) of Proposition 7.5.

Look at Figure 7.1. What sort of symmetry is indicated by the above functional equations, which imply that if $\zeta_X(u)$ has a pole at u then it must also have a pole at $1/(qu)$? If u is on the circle then $1/(qu)$ is the complex conjugate of u. If u is in the interval $\big(1/\sqrt{q},1\big)$ then $1/(qu)$ is in the interval $\big(1/q,1/\sqrt{q}\big)$.

To produce examples of regular graphs, the easiest method is to start with a generating set S of your favorite finite group G. Assume that S is **symmetric**, meaning that $s\in S$ implies $s^{-1}\in S$. Create a graph called a Cayley graph. It allows you to visualize the group. You can add colors and directions on the edges to get an even better picture but we won't do that. The name comes from the mathematician Arthur Cayley.

Definition 7.6 The **Cayley graph**, denoted $X(G,S)$, has as its vertices the elements of a finite group G and has edges between vertices g and gs for all $g\in G$ and $s\in S$. Here S is a symmetric generating set for G.

Cayley graphs are always regular with degree $|S|$. We want S to be a generating set because we want a connected graph. We want S to be symmetric for the graph to be undirected.

The cube is $X(\mathbb{F}_2^3, S)$, where $\mathbb{F}_2$ denotes the field with two elements, $\mathbb{F}_2^3$ is the additive group of 3-vectors with entries in this field, and

$$S = \left\{ \begin{bmatrix} 1 \\ 0 \\ 0 \end{bmatrix}, \begin{bmatrix} 0 \\ 1 \\ 0 \end{bmatrix}, \begin{bmatrix} 0 \\ 0 \\ 1 \end{bmatrix} \right\}.$$

Another example is the Paley graph, considered in Section 9.2 below.

A large number of Cayley graphs were considered in Terras [133]. One example is $X(\mathbb{F}_q^n, S)$, where S consists of solutions $x \in \mathbb{F}_q^n$ of the equation $x_1^2 + \cdots + x_n^2 = a$ for some $a \in \mathbb{F}_q$. We called such graphs "Euclidean." There are also "non-Euclidean" graphs associated with finite fields, in which the distance is replaced by a finite analog of the Poincaré distance in the upper half plane. The question of whether these Euclidean and non-Euclidean graphs are Ramanujan can be translated into a question about bounds on exponential sums. See Section 9.2 below.

More examples of regular graphs come from Lubotzky, Phillips, and Sarnak [79]. See Section 9.2 below and [35]. Mathematica will create "random" regular graphs with the command `X=RegularGraph[d,n]`, where d is the degree and n is the number of vertices.

Exercise 7.6 Consider examples of regular graphs such as those mentioned above and find out whether they are Ramanujan graphs. Then plot the poles of the Ihara zeta function. You might also look at the level spacings of the poles as in Figure 5.6.

Exercise 7.7 Show that if X is a non-bipartite k-regular graph with $k \geq 3$ then $1 = \Delta_X$, the g.c.d. of the prime lengths, from Definition 2.12.

Exercise 7.8 Consider the zeta function of the graph X obtained by removing one edge from the tetrahedron graph. Does the Ihara zeta function of X satisfy a functional equation of the form in Proposition 7.5 with $u \to u/q$ replaced by $u \to Ru$, where the radius of convergence R of the Ihara zeta is R_X from Definition 2.4?

8

Irregular graphs: what is the Riemann hypothesis?

Next let us consider irregular graphs (again unweighted and satisfying our usual hypotheses). Motoko Kotani and Toshikazu Sunada [72] proved the following theorem.

Theorem 8.1 (Kotani and Sunada) *Suppose that a graph X satisfies our usual hypotheses (see Section 2.1) and has vertices with maximum degree $q+1$ and minimum degree $p+1$.*

(1) Every pole u of $\zeta_X(u)$ satisfies $R_X \leq |u| \leq 1$, with R_X from Definition 2.4, and

$$q^{-1} \leq R_X \leq p^{-1}. \tag{8.1}$$

(2) Every non-real pole u of $\zeta_X(u)$ satisfies the inequality

$$q^{-1/2} \leq |u| \leq p^{-1/2}. \tag{8.2}$$

(3) The poles of ζ_X on the circle $|u| = R_X$ have the form $R_X e^{2\pi i a/\Delta_X}$, where $a = 1, \ldots, \Delta_X$. Here Δ_X is from Definition 2.12.

Proof We postpone the proof until Chapter 11 on the edge zeta function. □

Horton [57] gave examples of graphs such that $1/R_X$ is as close as desired to a given positive real number such as π or e.

Now let us define two constants associated with the graph X.

Definition 8.2 Let ρ_X and $\rho_{X'}$ be such that

$$\rho_X = \max\{|\lambda| \,|\, \lambda \in \text{Spectrum } A_X\},$$

$$\rho_X' = \max\{|\lambda| \,|\, \lambda \in \text{Spectrum } A_X,\ |\lambda| \neq \rho_X\}.$$

We will say that the **naive Ramanujan inequality** is

$$\rho'_X \leq 2\sqrt{\rho_X - 1}. \tag{8.3}$$

Lubotzky [78] defined X to be **Ramanujan** if

$$\rho'_X \leq \sigma_X, \tag{8.4}$$

where σ_X is the **spectral radius of the adjacency operator on the universal covering tree** of X. Recall that the spectral radius of the operator A is the supremum of $|\lambda|$ such that $A - \lambda I$ has no inverse (as a bounded linear operator on the tree). See Terras [132] for more information on spectral theory and also Terras [133]. The inequalities (8.3) and (8.4) both reduce to Definition 2.6 for connected regular graphs, by Proposition 7.2.

Definition 8.3 The **average degree of the vertices** of X is denoted $\overline{d_X}$.

Hoory [54] proved the following theorem.

Theorem 8.4 *We have*

$$2\sqrt{\overline{d_X} - 1} \leq \sigma_X.$$

Proof For the special case in which the graph is regular, the proof will essentially be given below when we prove the result of Alon and Boppana, Theorem 9.8. For the irregular case, the reader is referred to Hoory [54]. □

From Theorem 8.4 one has a criterion for a graph X to be Ramanujan in Lubotzky's sense: it need only satisfy the **Hoory inequality**

$$\rho'_X \leq 2\sqrt{\overline{d_X} - 1}. \tag{8.5}$$

To develop the Riemann hypothesis for irregular graphs, the natural change of variable is $u = R_X^s$ with R_X from Definition 2.4. All poles of $\zeta_X(u)$ are then located in the "critical strip" $0 \leq \text{Re } s \leq 1$ with poles at $s = 0$ ($u = 1$) and $s = 1$ ($u = R_X$). The examples below show that, for irregular graphs, one cannot expect a functional equation relating $f(s) \equiv \zeta\left(R_X^s, X\right)$ and $f(1 - s)$. Therefore it is natural to say that the RH for X should require that $\zeta_X(u)$ has no poles in the open strip $1/2 < \text{Re } s < 1$. This is the graph theory RH below. After looking at examples, it seems that one rarely sees an Ihara zeta satisfying this RH (although random graphs do seem to satisfy the RH approximately). Thus, below we also consider the weak graph theory RH.

The **graph theory RH** says that $\zeta_X(u)$ is pole free for

$$R_X < |u| < \sqrt{R_X}. \tag{8.6}$$

The **weak graph theory RH** says that $\zeta_X(u)$ is pole free for

$$R_X < |u| < 1/\sqrt{q}. \tag{8.7}$$

Note that (8.6) and (8.7) are the same if the graph is regular. We have examples (such as Example 8.5 below) for which $R_X > q^{-1/2}$ and in such cases the weak graph theory RH is true but vacuous. In Stark and Terras [122] we gave a longer discussion of the preceding two versions of the RH for graphs showing the connections with the versions of the RH for the Dedekind zeta function and the existence of Siegel zeros. We will define Siegel zeros for the Ihara zeta function later.

Sometimes number theorists state a modified generalized Riemann hypothesis (GRH) for the Dedekind zeta function and this just ignores all possible real zeros while only requiring the non-real zeros to be on the line Re $s = 1/2$. The graph theory analog of the modified weak GRH would just ignore the real poles and require that there are no non-real poles of $\zeta_X(u)$ in $R_X < |u| < q^{-1/2}$. But this is true for all graphs by Theorem 8.1: if μ is a pole of $\zeta_X(u)$ and $|\mu| < q^{-1/2}$ then μ is real!

One may ask about the relations between the constants $\rho_X, \overline{d_X}, R_X$. One can show (**exercise**) that

$$\rho_X \geq \overline{d_X}. \tag{8.8}$$

This is easily seen using the fact that ρ_X is the maximum value of the Rayleigh quotient $\langle Af, f\rangle / \langle f, f\rangle$, where A is an adjacency matrix and f is any non-zero vector in $\mathbb{R}^n$, while $\overline{d_X}$ is the value when f is the vector all of whose entries are 1. In all the examples to date we can see that

$$\rho_X \geq 1 + \frac{1}{R_X} \geq \overline{d_X} \tag{8.9}$$

but can only show (see Proposition 11.17 in Section 11.2) that

$$\rho_X \geq \frac{p}{q} + \frac{1}{R_X}.$$

As a **research problem**, the reader might want to investigate possible improvements to the last formula.

Next we give some examples including answers to the following questions: Do the spectra of the adjacency matrices satisfy the naive Ramanujan inequality (8.3) or the Hoory inequality (8.5)? Do the Ihara zeta functions for the graphs have the pole-free region (8.7) of the weak graph theory RH or the pole-free region (8.6) of the full graph theory RH?

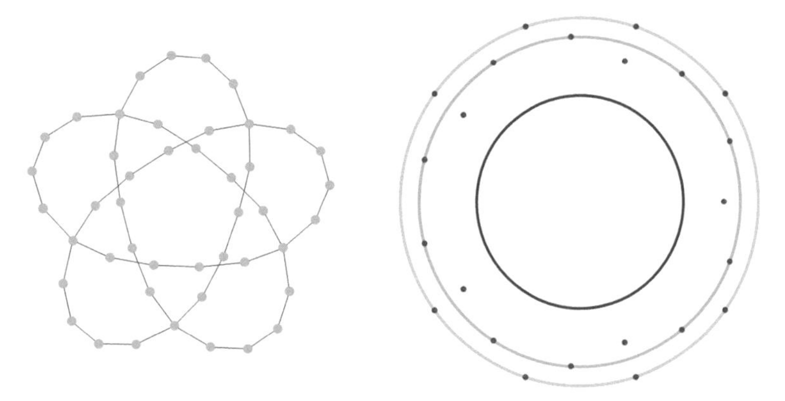

Figure 8.1 *Left,* the graph $X = Y_5$ is obtained by adding four vertices to each edge of $Y = K_5$. *Right,* the poles ($\neq -1$) of the Ihara zeta function of X are the magenta points. The circles have centers at the origin and radii $\{q^{-1/2}, R^{1/2}, p^{-1/2}\}$. Note the five-fold rotational symmetry of the poles.

Example 8.5 Let X be the **graph obtained from the complete graph on five vertices by adding four vertices to each edge**, as shown on the left in Figure 8.1.

For the graph X, we find that $\rho' \approx 2.327\,71$ and

$$\{\rho, 1 + 1/R, \overline{d_X}\} \approx \{2.391\,38, 2.245\,73, 2.222\,22\}.$$

This graph satisfies the naive Ramanujan inequality (8.3) but not the Hoory inequality (8.5). The magenta points on the right in Figure 8.1 are the poles of $\zeta_X(u)$ not equal to -1. Here

$$\zeta_X(u)^{-1} = \zeta_{K_5}(u^5)^{-1} = (1 - u^{10})^5 (1 - 3u^5)(1 - u^5)(1 + u^5 + 3u^{10}),$$

where K_5 is the complete graph on five vertices. The circles on the right are centered at the origin, with radii

$$\{q^{-1/2}, R, R^{1/2}, p^{-1/2}\} \approx \{0.57735, 0.802742, 0.895958, 1\}.$$

The zeta function satisfies the RH and thus the weak RH. However, the weak RH is vacuous in this case.

Example 8.6: Random graph with probability 1/2 of an edge The magenta points in Figure 8.2 are the poles not equal to ± 1 of the Ihara zeta function of a random graph produced by Mathematica with the command `RandomGraph[100,1/2]`. This means that there are 100 vertices and that the probability of an edge between any two vertices is $1/2$. The graph satisfies

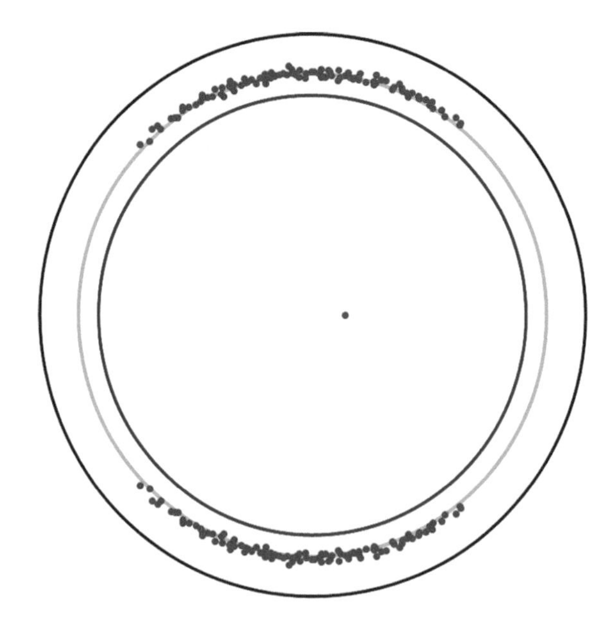

Figure 8.2 The magenta points are poles ($\neq \pm 1$) of the Ihara zeta function for a random graph, produced by Mathematica with the command `RandomGraph[100,1/2]`. The circles have centers at the origin and radii $\{q^{-1/2}, R^{1/2}, p^{-1/2}\}$. The RH looks approximately true, but in fact it is not exactly true. The weak RH is true.

the Hoory inequality (8.5) and it is thus Ramanujan in Lubotzky's sense. It also satisfies the naive Ramanujan inequality (8.3). We find that $\rho' \approx 10.0106$ and $\{\rho, 1 + 1/R, \overline{d_X}\} \approx \{50.054, 50.0435, 49.52\}$. The circles in Figure 8.2 are centered at the origin and have radii given by $\{q^{-1/2}, R^{1/2}, p^{-1/2}\} \approx \{0.130\,189, 0.142\,794, 0.166\,667\}$. The poles of the zeta function satisfy the weak RH but not the RH. However, the RH seems to be approximately true. See Skiena [114] for more information on the model that Mathematica uses to produce random graphs.

Example 8.7: Torus minus some edges From the torus graph T which is the product of a 10-cycle and a 20-cycle, we delete six edges to obtain a graph that we will call N, shown on the left in Figure 8.3. The spectrum of the adjacency matrix of N satisfies neither the Hoory inequality (8.5) nor the naive Ramanujan inequality (8.3). We find that $\{\rho, 1 + 1/R, \overline{d}\} \approx \{3.987\,49, 3.985\,68, 3.98\}$ and $\rho' \approx 3.902\,75$. The right-hand side of Figure 8.3 shows the poles of the Ihara zeta for N as magenta points. The circles are centered at the

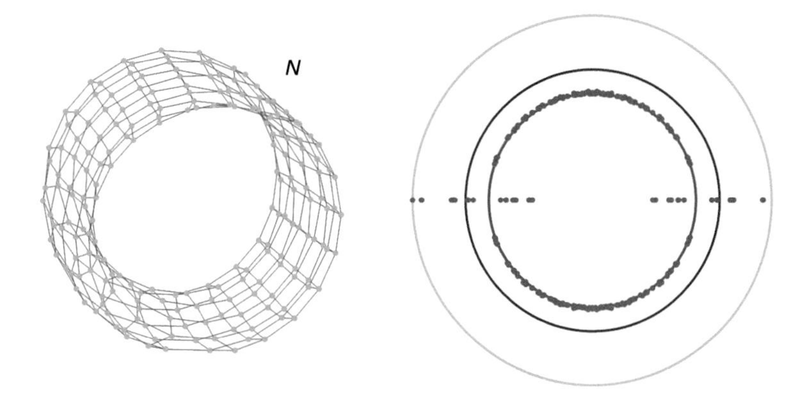

Figure 8.3 *Left,* the graph N results from deleting six edges from the product of a 10-cycle and a 20-cycle (a torus). *Right,* the magenta points indicate the poles ($\neq \pm 1$) of the Ihara zeta function of N. The circles are centered at the origin, with radii $\{1/\sqrt{q}, \sqrt{R}, 1/\sqrt{p}\}$. The Ihara zeta function satisfies neither the RH nor the weak RH.

origin and have radii $\{q^{-1/2}, R^{1/2}, p^{-1/2}\} \approx \{0.577\,35, 0.578\,73, 0.707\,11\}$. The zeta poles satisfy neither the graph theory weak RH nor the RH.

Exercise 8.1 Consider the graph $X = K_n - e$, the complete graph on n vertices, K_n, with one edge e removed.

(a) Does the zeta function for X satisfy the RH?
(b) Does it satisfy the weak RH?
(c) Does X satisfy the naive Ramanujan inequality (8.3)?
(d) Does X satisfy the Hoory inequality (8.5)?
(e) Is there a functional equation relating $\zeta_X(u)$ and $\zeta_X(R_X/u)$?

Hint: See [122], where most of these questions are answered explicitly.

Example 8.9 Figures 8.4 and 8.5 show the results of some Mathematica experiments on the distribution of the poles of the Ihara zeta for two graphs. The graphs were constructed using the `RealizeDegreeSequence` command as well as the commands `GraphUnion` and `Contract`. In both figures, the top diagram shows the graph. The middle diagram shows the histogram of degrees. The magenta points in the bottom diagram are the poles of the Ihara zeta corresponding to the graph in the top diagram. Many of these poles violate the RH by being inside the green circle rather than outside. Those poles violating the weak RH are inside the inner circle; no such poles occur for the graph in Figure 8.5, but there are such poles for the graph in Figure 8.4.

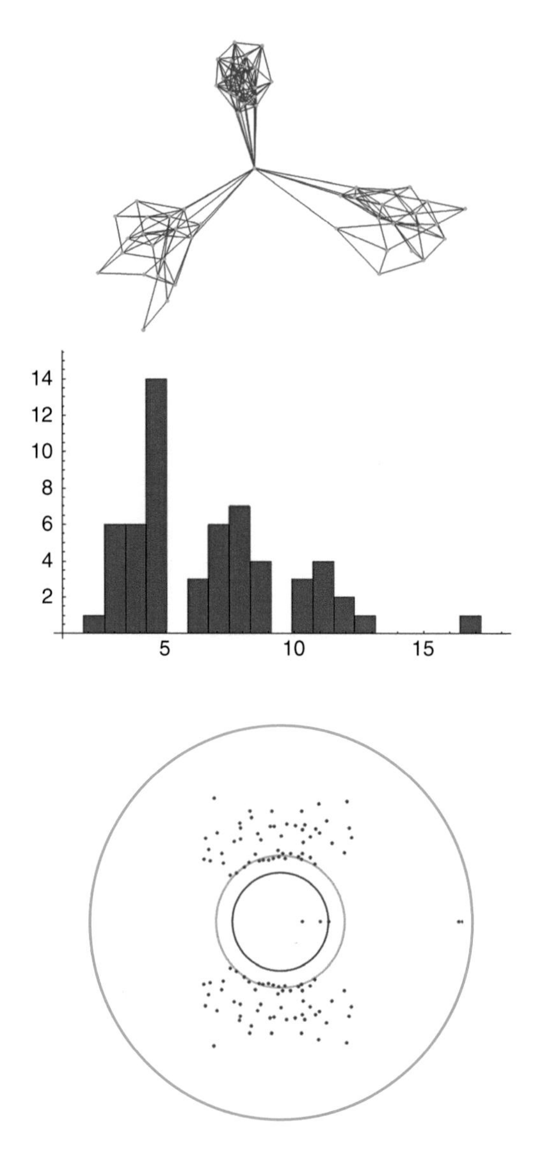

Figure 8.4 A Mathematica experiment. *Top*, the graph; *middle*, the histogram of degrees; *bottom*, the poles, shown in magenta, of the Ihara zeta function of the graph. The intermediate circle, shown in green, is the RH circle, with radius $\sqrt{R}$, where R is the closest pole to 0. The smallest circle has radius $1/\sqrt{q}$, where $q + 1$ is the maximum degree of the graph. The largest circle has radius 1. For this graph $p = 1$ and thus the circle of radius $1/\sqrt{p}$ coincides with the circle of radius 1. Many poles are inside the green middle circle and thus violate the RH. For this graph, the RH and the weak RH are false, as is the naive Ramanujan inequality.
The probability of an edge is 0.119 177.

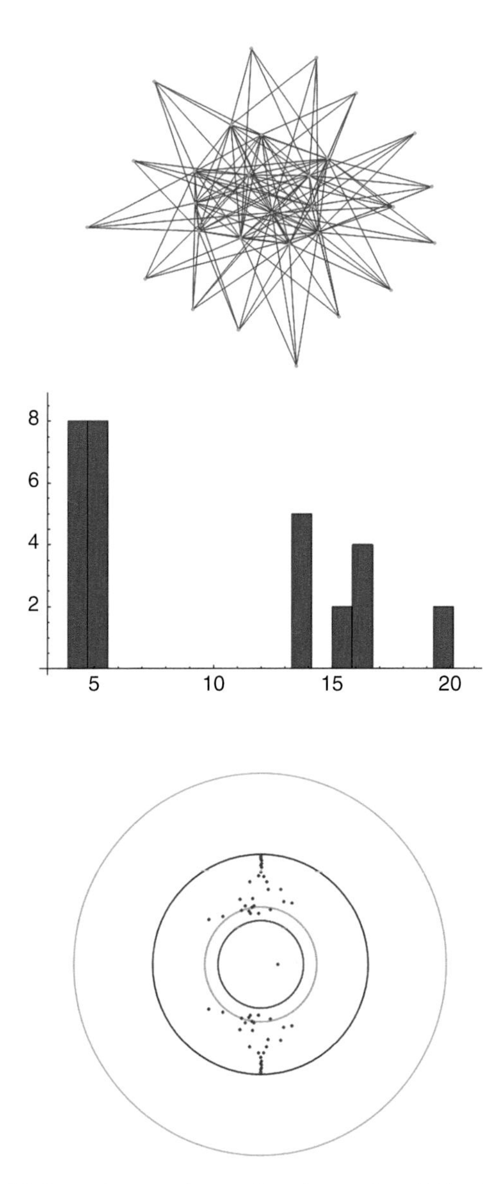

Figure 8.5 A Mathematica experiment. *Top*, the graph; *middle*, the histogram of degrees; *bottom*, the poles, shown in magenta, of the Ihara zeta function of the corresponding graph. The smallest circle has radius $1/\sqrt{q}$, where $q + 1$ is the maximum degree of the graph. The next circle out (the green circle) is the RH circle with radius $\sqrt{R}$, where R is the closest pole to 0. The largest circle has radius 1. The circle just inside this has radius $1/\sqrt{p}$, where $p + 1$ is the minimum degree of the graph. For this graph, the RH is false, but the weak RH is true as well as the naive Ramanujan inequality. The probability of an edge is 0.339 901 for this graph.

Exercise 8.2 Do more examples in the spirit of the preceding figures. The pole locations in Figures 8.2 and 8.3 do not appear too different from those for a regular graph in Figure 7.1. The easiest way to see a more two-dimensonal pole picture would be to use the command `RandomGraph[100,.1]`. There are many other ways to construct new graphs in Mathematica, for example, those used to obtain Figures 8.4 and 8.5, where we employed the `RealizeDegreeSequence` command as well as the commands `GraphUnion` and `Contract`. There are also many examples of covering graphs in later chapters of this book. Or you could construct some zig-zag products (see Hoory, Linial, and Wigderson [55]). If you have access to a humongous computer, try the internet graph.

9
Discussion of regular Ramanujan graphs

In this chapter we restrict ourselves to regular graphs. Our goals are:

(1) to explain why a random walker gets lost fast on a Ramanujan graph;
(2) to give examples of regular Ramanujan graphs;
(3) to show why the Ramanujan bound is best possible;
(4) to explain why Ramanujan graphs are good expanders;
(5) to give a diameter bound for a Ramanujan graph.

9.1 Random walks on regular graphs

Suppose that A is the adjacency matrix of a k-regular graph X with n vertices. We obtain a **Markov chain** from A as follows. The states are the vertices of X. At time t, the process (walker) goes from the ith state to the jth state, with probability p_{ij} given by $1/k$ if vertex i is adjacent to vertex j and by 0 otherwise. A **probability vector** $p \in \mathbb{R}^n$ has non-negative entries p_i such that $\sum_{i=1}^n p_i = 1$. Here p_i represents the probability that the random walker is at vertex i of the graph.

Notation 9.1 All our vectors in $\mathbb{R}^n$ are column vectors and, as before, we write ${}^{\mathrm{t}}p$ to denote the **transpose** of a column vector p in $\mathbb{R}^n$. The same notation will also be used for matrices.

The **Markov transition matrix** is

$$T = (p_{ij})_{1 \le i,j \le n} = \frac{1}{k}A.$$

Let $p_i^{(m)}$ denote the probability that the walker is at vertex i at time m. The **probability vector** is $p^{(m)} = {}^{\mathrm{t}}\big(p_1^{(m)} \;\cdots\; p_n^{(m)}\big)$. Then

$$p^{(m+1)} = Tp^{(m)} \qquad \text{and} \qquad p^{(m)} = T^m p^{(0)}.$$

Theorem 9.2 (A random walker gets lost) *Suppose that X is a connected non-bipartite k-regular graph with n vertices and adjacency matrix A. If $T = (1/k)A$, for every initial probability vector $p^{(0)}$ we have*

$$\lim_{m\to\infty} p^{(m)} = \lim_{m\to\infty} T^m p^{(0)} = u = {}^{\mathrm{t}}\left(\frac{1}{n} \cdots \frac{1}{n}\right);$$

i.e., the limit is the uniform probability vector.

Proof Since T is a real symmetric matrix, the spectral theorem from linear algebra says that there is a real orthogonal matrix U, i.e., ${}^{\mathrm{t}}UU = I$, the identity matrix, such that ${}^{\mathrm{t}}UTU = D$, where D is a diagonal matrix with the eigenvalues λ_i of T down the diagonal. Let $U = (u_1 \ \cdots \ u_n)$, with column vectors u_i. Then these columns are orthonormal, meaning that the inner products $\langle u_i, \ u_j\rangle$ of the columns satisfy

$$\langle u_i, u_j\rangle = {}^{\mathrm{t}}u_i u_j = \begin{cases} 1, & i = j, \\ 0, & i \neq j. \end{cases}$$

Now any vector v can be written as a linear combination of the column vectors of U:

$$v = \sum_{i=1}^{n} \langle v, u_i\rangle u_i \qquad \text{and} \qquad Tv = \sum_{i=1}^{n} \langle v, u_i\rangle \lambda_i u_i.$$

Then

$$T^m v = \sum_{i=1}^{n} \langle v, u_i\rangle \lambda_i^m u_i.$$

Assume that u_1 is a constant vector of norm 1 (with eigenvalue $\lambda_1 = 1$). Let the entries of u_1 be $1/\sqrt{n}$. By the hypothesis on X we know (from Proposition 7.2) that $|\lambda_i| < 1$ for $i > 1$. It follows that

$$\lim_{m\to\infty} \lambda_i^m = \begin{cases} 1, & i = 1, \\ 0, & i \neq 1. \end{cases}$$

Thus

$$\lim_{m\to\infty} T^m v = \langle v, u_1\rangle u_1 = \frac{1}{n}.$$

This proves the theorem. □

We want to know how long it takes the random walker to get lost. This depends on the second largest eigenvalue in absolute value of the adjacency matrix, assuming that the graph is non-bipartite. The next theorem answers the question. If the graph is bipartite, one can modify the random walk to make the walker get lost, by allowing the walker to stay in the same place with equal probability. We will use the 1-norm $\| \ \|_1$ to measure distances between vectors in $\mathbb{R}^n$. Statisticians seem to prefer this to the 2-norm. See Diaconis [36]. Define

$$\|v\|_1 = \sum_{i=1}^{n} |v_i| \,. \tag{9.1}$$

Theorem 9.3 (How long does it take to get lost?) *Suppose that X is a connected non-bipartite k-regular graph with n vertices and adjacency matrix A. If $T = (1/k)A$, for every initial probability vector $p^{(0)}$, we have*

$$\left\| T^m p^{(0)} - u \right\|_1 \leq \sqrt{n} \left(\frac{\mu}{k} \right)^n ,$$

where

$$u = {}^{\mathrm{t}}\left(\frac{1}{n}, \ldots, \frac{1}{n} \right)$$

and

$$\mu = \max\left\{ |\lambda| \middle| \lambda \in \text{Spectrum } A, |\lambda| \neq q+1 \right\}.^1$$

Proof See Terras [133], pp. 104–106. The proof is in the same spirit as that of the preceding theorem. □

Corollary 9.4 *If the graph in Theorem 9.3 is Ramanujan (see Definition 2.6) then $\mu \leq 2\sqrt{k-1}$ and*

$$\left\| T^m p^{(0)} - u \right\|_1 \leq \sqrt{n} \left(\frac{2\sqrt{k-1}}{k} \right)^n .$$

The moral of this story is that for moderately large values of the degree k it does not take a very long time before the walker is lost.

Exercise 9.1 Redo the preceding results for irregular graphs.

9.2 Examples: the Paley graph, two-dimensional Euclidean graphs, and the graphs of Lubotzky, Phillips, and Sarnak

We consider various examples, all of which are Cayley graphs $X(G, S)$ from Definition 7.6. When G is the cyclic group $\mathbb{Z}_n = \mathbb{Z}/n\mathbb{Z}$, its **characters** are of the form $\chi_a(y) = \exp(2\pi i \, ay/n)$, where $a, y \in G$. These form a basis for the

[1] The constant μ equals ρ' from Definition 8.2.

eigenfunctions of the adjacency matrix A viewed as an operator on functions $f : G \to \mathbb{C}$ via the formula

$$Af(y) = \sum_{x \in S} f(x+y),$$

for $x \in G$. That is,

$$A\chi_a = \lambda_a \chi_a, \text{ where } \lambda_a = \sum_{x \in S} \chi_a(x). \tag{9.2}$$

See Terras [133] for more information on this subject.

Exercise 9.2 Prove formula (9.2).

Example 9.5: The Paley graph $P = X(\mathbb{Z}/p\mathbb{Z}, \square)$ is a Cayley graph for the group $\mathbb{Z}/p\mathbb{Z}$, where p is an odd prime of the form $p = 1 + 4n,\ n \in \mathbb{Z}$.[2] The vertices of the graph are elements of $\mathbb{Z}/p\mathbb{Z}$, and two vertices a, b are connected iff $a - b$ is a non-zero square in $\mathbb{Z}/p\mathbb{Z}$. If $p \equiv 1 (\mathrm{mod}\, 4)$ then -1 is a square (mod p), and the converse also holds (**exercise**). It follows that when $p \equiv 1 (\mathrm{mod}\, 4)$ the Paley graph is undirected.

The **characters** of the group $\mathbb{Z}/p\mathbb{Z}$ are of the form $\chi_a(y) = \exp(2\pi i a y/p)$, for $a, y \in \mathbb{Z}/p\mathbb{Z}$. They form a complete orthogonal set of eigenfunctions of the adjacency operator of the Paley graph. We have

$$A\chi_a(y) = \sum_{x \sim y} \exp \frac{2\pi i a x}{p} = \frac{1}{2} \sum_{\substack{x = y + u^2 \\ 0 \neq u \in \mathbb{Z}/p\mathbb{Z}}} \exp \frac{2\pi i a x}{p} = \lambda_a \chi_a(y).$$

The eigenvalues λ_a have the form

$$\lambda_a = \frac{1}{2} \sum_{u=1}^{p-1} \exp \frac{2\pi i a u^2}{p}.$$

Recall that the **Gauss sum** is

$$G_a = \sum_{u=0}^{p-1} \exp \frac{2\pi i a u^2}{p}. \tag{9.3}$$

Thus $\lambda_a = (G_a - 1)/2$. Use the exercise below to see that if a is not congruent to $0 (\mathrm{mod}\, p)$ then

$$|\lambda_a| \le \frac{1 + \sqrt{p}}{2}.$$

Thus the graph is Ramanujan if $p \ge 5$, since the degree is $(p-1)/2$.

[2] Paley and Cayley are two different mathematicians. The Paley graph is a special case of a Cayley graph.

Exercise 9.3 Show that when a is not congruent to $0 (\mathrm{mod}\, p)$, the Gauss sum defined in the preceding example satisfies $|G_a| = \sqrt{p}$.

Hint: This can be found in most elementary number theory books and in Terras [133].

Exercise 9.4 Fill in the details showing that the Paley graph $P = X(\mathbb{Z}/p\mathbb{Z}, \square)$ above is Ramanujan when $p \geq 5$. Then find out how large p must be in order that $\|T^m v - u\|_1 \leq 1/100$, where

$$v = {}^{\mathrm{t}}(1\,0\,0 \cdots 0) \qquad \text{and} \qquad u = {}^{\mathrm{t}}(1/p \cdots 1/p).$$

Example 9.6: Two-dimensional Euclidean graphs Suppose that p is an odd prime. Define the Cayley graph $X(G, S)$ for the group $G = \mathbb{F}_p^2$ consisting of 2-vectors with entries in $\mathbb{F}_p = \mathbb{Z}/p\mathbb{Z}$, with the operation of vector addition. The generating set S is the set of vectors

$$\begin{pmatrix} x \\ y \end{pmatrix} \in \mathbb{F}_p^2$$

satisfying $x^2 + y^2 = 1$. This is a special case of the Euclidean graphs considered in Terras [133], where they are connected with finite analogs of symmetric spaces.

The characters of $G = \mathbb{F}_p^2$ are

$$\psi_{a,b}\begin{pmatrix} x \\ y \end{pmatrix} = \exp\left(\frac{2\pi i\,(ax + by)}{p}\right),$$

for (a, b) and $(x, y) \in G$. They form a complete orthogonal set of eigenfunctions of the adjacency operator of the two-dimensional (2D) Euclidean graph:

$$\begin{aligned} A\psi_{a,b}\begin{pmatrix} u \\ v \end{pmatrix} &= \sum_{\binom{x}{y} \sim \binom{u}{v}} \exp\left(\frac{2\pi i ax + by}{p}\right) \\ &= \sum_{r^2+s^2=1} \exp\left(\frac{2\pi i\,(a\,(r+u) + b\,(s+v))}{p}\right) = \lambda_{a,b}\psi_{a,b}\begin{pmatrix} u \\ v \end{pmatrix}. \end{aligned}$$

The corresponding eigenvalues $\lambda_{a,b}$ are given by

$$\lambda_{a,b} = \sum_{r^2+s^2=1} \exp\left(\frac{2\pi i\,(ar + bs)}{p}\right).$$

These numbers can be identified with a sum which is a favorite of number theorists, called a Kloosterman sum.

If κ is a character of the multiplicative group $\mathbb{F}_p^* = \mathbb{F}_p - 0$ and $a, b \in \mathbb{F}_p^*$, define the **generalized Kloosterman sum** as

$$K(\kappa|a, b) = \sum_{t \in \mathbb{F}_p^*} \kappa(t) \exp \frac{-2\pi i\,(at + b/t)}{p}.$$

It turns out that the non-trivial eigenvalues of the adjacency matrix for the 2D Euclidean Cayley graph are

$$\lambda_{a,b} = \frac{1}{p} G_1^2 K\left(\varepsilon^2 \middle| 1, a^2 + b^2\right),$$

where G_1 is the Gauss sum in formula (9.3) and the quadratic character is

$$\varepsilon(t) = \begin{cases} 1, & t \equiv u^2 (\mathrm{mod}\, p) \text{ for some } u \in \mathbb{F}_p^*, \\ 0, & t \equiv 0 (\mathrm{mod}\, p), \\ -1 & \text{otherwise.} \end{cases} \tag{9.4}$$

As a consequence of the Riemann hypothesis for zeta functions of curves over finite fields one has a bound on the Kloosterman sums. This was proved by A. Weil. See Rosen [104] for more information. The bound implies that for $(a, b) \neq (0, 0)$ we have

$$\left|\lambda_{a,b}\right| \leq 2\sqrt{q}.$$

The degrees of the two-dimensional Euclidean Cayley graph may be computed exactly to be $p - \varepsilon(-1)$, where ε is defined by formula (9.4). See Rosen [104].

It follows that if $p \equiv 3 (\mathrm{mod}\, 4)$, the graphs are Ramanujan. But when $p = 17$ or 53, for example, the graphs are not Ramanujan. Katz has proved that these Kloosterman sums do not vanish. He also proved that the distribution of the Kloosterman sums approaches the semicircle distribution as $p \to \infty$. See Terras [133] for references. Newland [95] found that, for these Euclidean graphs, the level spacings of Im s corresponding to the poles of $\zeta(q^{-s}, X)$ look Poisson for large p. The contour maps of the eigenfunctions are beautiful pictures of the finite circles $x^2 + y^2 \equiv a (\mathrm{mod}\, p)$. There are movies on my website of these pictures as p runs through an increasing sequence of primes.

In Terras [133] we also define non-Euclidean finite upper half plane graphs where the Euclidean distance is replaced by a finite analog of the Poincaré distance. These also give Ramanujan graphs, the adjacency matrices of which provide interesting spectra. It takes quite a bit of knowledge of group representations plus Weil's result proving the Riemann hypothesis for curves over finite fields in order to prove that the finite upper half plane graphs are Ramanujan.

Exercise 9.5 Consider some examples of the finite upper half plane graphs in [133]. Experiment with the spectra of the adjacency matrices to see whether the graphs are Ramanujan. Look at the level spacings of the poles of the Ihara zeta, or replace squares by fourth powers in the Paley or Euclidean graphs.

Such examples are not really the expander graphs sought by computer scientists since the degree blows up with the number of vertices. It is more difficult to find families of Ramanujan graphs of fixed degree as the number of vertices approaches infinity. The first examples were due to Margulis [82] and independently Lubotzky, Phillips, and Sarnak [79]. Friedman [42] proved that for fixed degree k and and $\epsilon > 0$ the probability that $\lambda_1(X_{m,k}) \leq 2\sqrt{k-1}+\epsilon$ approaches 1 as $n \to \infty$. Miller, Novikoff and Sabelli [87] gave evidence for the following conjecture: the probability that a regular graph is exactly Ramanujan is approximately 27%.

We finish the chapter by presenting the example of Lubotzky *et al.* [79].

Example 9.7: Lubotzky, Phillips, and Sarnak graphs $X_{p,q}$ Let p and q be distinct primes congruent to $1 \pmod 4$. The graphs $X_{p,q}$ are Cayley graphs for the group $G = \mathrm{PGL}(2, \mathbb{F}_q) = \mathrm{GL}(2, \mathbb{F}_q)/Center$. Here $\mathrm{GL}(2, \mathbb{F}_q)$ is the group of non-singular 2×2 matrices with elements in the field having q elements and *Center* consists of matrices which are non-zero scalar multiples of the identity.

Fix some integer i such that $i^2 \equiv -1 \pmod q$. Define S to be

$$S = \left\{ \begin{pmatrix} a_o + ia_1 & a_2 + ia_3 \\ -a_2 + ia_3 & a_0 - ia_1 \end{pmatrix} \middle| \, a_0^2 + a_1^2 + a_2^2 + a_3^2 = p \right.$$
$$\left. \text{for odd } a_0 > 0 \text{ and even } a_1, a_2, a_3 \right\}.$$

A theorem of Jacobi says that there are exactly $p+1$ integer solutions to $a_0^2+a_1^2+a_2^2+a_3^2=p$, so that $|S|=p+1$. One can show that S is closed under matrix inversion. The graph $X_{p,q}$ is then the connected component of the identity in the Cayley graph $X(G, S)$. It can be proved that either $X(G, S)$ is connected or it has two connected components of equal size. Using Weil's proof of the Riemann hypothesis for zeta functions of curves over finite fields, Lubotzky, Phillips, and Sarnak showed that these graphs are Ramanujan. For fixed p we then have a family of Ramanujan graphs of degree $p+1$ having $O(q^3)$ vertices as $q \to \infty$.

Exercise 9.6 Compute the Ihara zeta functions for some of the graphs in this section.

9.3 Why the Ramanujan bound is best possible (Alon and Boppana theorem)

We want to prove the following theorem.

Theorem 9.8 (Alon and Boppana) *Suppose that X_n is a sequence of k-regular connected graphs and that the number of vertices of X_n approaches infinity with n. Let $\lambda_1(X_n)$ denote the second largest eigenvalue of the adjacency matrix of X_n. Then*

$$\lim_{n\to\infty} \inf_{m\geq n} \lambda_1(X_m) \geq 2\sqrt{k-1}.$$

Proof (Lubotzky, Phillips, and Sarnak) Let the set of eigenvalues of the adjacency matrix A_n of X_n be

$$\text{Spec } A_n = \left\{\lambda_0 = k > \lambda_1 \geq \lambda_2 \geq \cdots \geq \lambda_{|V(X_n)|-1}\right\}.$$

Let $N_v(m, X_n)$ be the number of paths of length m going from vertex v and back to v in graph X_n. Note that these paths can have backtracking and tails. Then

$$\sum_{j=0}^{|V(X_n)|-1} \lambda_j^m = \text{Tr } A_n^m = \sum_{v\in X} N_v(m, X_n).$$

The universal covering space of X_n is the k**-regular tree** T_k (meaning that it is an infinite graph which is k-regular, connected, and has no cycles). Part of T_4 is pictured in Figure 2.4. The lower bound we seek is actually the spectral radius of the adjacency operator on T_k.

Let τ_m be the number of paths of length m on T_k going from any vertex $\widetilde{v}$ back to $\widetilde{v}$. Since T_k is the k-regular tree, τ_m is 0 unless m is even and τ_m is independent of $\widetilde{v}$. Then $N_v(m, X_n) \geq \tau_m$, since any path on T_k projects down to a path on X_n. Therefore

$$\sum_{j=0}^{|V(X_n)|-1} \lambda_j^m = \sum_{v\in X_n} N_v(m, X_n) \geq |V(X_n)|\, \tau_m.$$

It follows that

$$k^m + (|V(X_n)| - 1)\, \lambda_1^m \geq |V(X_n)|\, \tau_m.$$

We will be done if we can show that

$$\tau_{2m}^{1/2m} \to 2\sqrt{k-1} \qquad \text{as } m \to \infty, \tag{9.5}$$

for then we would have

$$\lambda_1 \geq \left(\frac{|V(X_n)|\,\tau_{2m} - k^{2m}}{|V(X_n)| - 1} \right)^{1/2m}$$

$$= \left(\frac{|V(X_n)|}{|V(X_n)| - 1} \right)^{1/2m} \tau_{2m}^{1/2m} \left(1 - \frac{k^{2m}}{\tau_{2m}\,|V(X_n)|} \right)^{1/2m}.$$

The first factor on the right-hand side of the equality approaches 1 as $n \to \infty$. The second factor approaches $2\sqrt{k-1}$. The third factor approaches 1.

Now we must prove formula (9.5). For this part of the proof we follow the reasoning of H. Stark. Let x and y be any two points of T_k such that the distance between them $d(x, y)$ equals j, i.e., the number of edges in the unique path in T_k joining x and y is j.

We define $\tau(n, j)$ to be the number of ways starting at x to get to a point y at distance j from x by a path of length n in T_k. It is $\tau(m, 0) \equiv \tau_m$ that we want to study. It is an **exercise** to see that $\tau(n, j) \neq 0$ implies that $j \equiv n (\text{mod}\, 2)$ and that $n \geq j$. For $j > 0$ and $n > 1$, we have the recursion

$$\tau(n, j) = (k-1)\tau(n-1, j+1) + \tau(n-1, j-1). \tag{9.6}$$

The reason is that we must be at one of the k neighbors of y at the $(n-1)$th step, and $k-1$ of these neighbors are a distance $j+1$ from x while the last neighbor is at a distance $j-1$ from x. The recursion (9.6) is reminiscent of Pascal's triangle. It is an **exercise** to show that

$$\tau(2m, 0) \geq a(2m, 0)(k-1)^m,$$

where $a(n, j)$ is defined by the following recursive definition:

$$a(n, j) = a(n-1, j-1) + a(n-1, j+1);$$

$$a(0, 0) = 1;$$

$$a(n, 0) = 0 \qquad \text{unless } 0 \leq j \leq n,\ a(0, 0) = 1.$$

Set $a_{2m} = a(2m, 0)$. Note that a_{2m} satisfies the recursion

$$a_{2m} = \sum_{k=1}^{m} a_{2k-2}a_{2m-2k}.$$

This recursion arises in many ways in combinatorics; see Vilenkin [140]. For example, a_{2m} is the number of permutations of $2m$ letters, m of which are b's and m of which are f's, such that for every r with $1 \leq r \leq 2m$ the number of b's in the first r terms of the permutations is greater than or equal to the

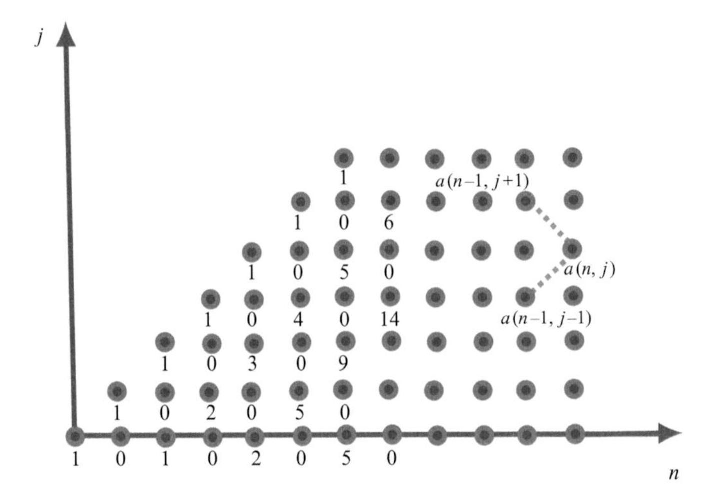

Figure 9.1 Part of the proof of Theorem 9.8. The values of $a(n, j)$ are defined by the recursion $a(n, j) = a(n-1, j-1) + a(n-1, j+1)$, with $a(0, 0) = 1$ and $a(n, j) = 0$ unless $0 \leq j \leq n$.

number of f's. The solution is the **Catalan number**

$$a_{2m} = \frac{1}{m+1}\binom{2m}{m}.$$

Stirling's formula implies that $\binom{2m}{m}^{1/2m} \sim 2$ as $m \to \infty$. Formula (9.5) follows and thus the theorem. □

Exercise 9.7 Fill in the details in the preceding proof. It may help to look at Figure 9.1 showing some of the values of $a(n, j)$.

9.4 Why are Ramanujan graphs good expanders?

First, what is an expander graph? Roughly it means that the graph is highly connected but sparse (there are relatively few edges). Such graphs are useful in computer science – for building efficient communication networks and for creating error-correcting codes with efficient encoding and decoding. See Giuliana Davidoff *et al.* [35], Hoory, Linial, and Wigderson [55], Lubotzky [77], or Sarnak [109] for more information. Fan Chung [27] provides a discussion of expansion in irregular graphs.

Suppose that X is an undirected k-regular graph satisfying our usual assumptions.

Definition 9.9 For sets of vertices S, T of X, define

$$E(S,T) = \{e|e \text{ is an edge of } X \text{ with one vertex in } S \text{ and the other vertex in } T\}.$$

Definition 9.10 If S is a set of vertices of X, we say the **boundary** is $\partial S = E(S, X - S)$.

Definition 9.11 A graph X with vertex set V and $n = |V|$ has **expansion ratio**

$$h(X) = \min_{\{S \subset V \,||S| \leq n/2\}} \frac{|\partial S|}{|S|}.$$

Note that there are many variations on this definition. We follow Sarnak [109] and Hoory *et al.* [55] here. The expansion constant is the discrete analog of the Cheeger constant in differential geometry. See Lubotzky [77].

Definition 9.12 A sequence of $(q+1)$-regular graphs $\{X_j\}$ such that $|V(X_j)| \to \infty$ as $j \to \infty$ is called an **expander family** if there is an $\varepsilon > 0$ such that the expansion ratio $h(X_j) \geq \varepsilon$ for all j.

For connected k-regular graphs X whose adjacency matrix has the spectrum $k = \lambda_1 > \lambda_2 \geq \cdots \geq \lambda_n$ one can prove that

$$\frac{k - \lambda_2}{2k} \leq h(X). \tag{9.7}$$

See Terras [133], p. 80. There is also a discussion in Hoory, Lineal, and Wigderson [55], pp. 474–476, where an upper bound was proved as well. Such results were originally proved by Dodziuk, Alon, and Milman. It follows that for a large expansion constant, one needs a small λ_2.

Next we prove the expander-mixing lemma (from Alon and Fan Chung [1]) which implies that $E(S,T)$ will be closer to the expected number of edges between S and T in a random k-regular graph X of edge density k/n (where $n = |V|$) provided that μ, the second largest eigenvalue (in absolute value) of the adjacency matrix of X is small as possible.

Lemma 9.13 (Expander-mixing lemma) *Suppose that X is a connected k-regular non-bipartite graph with n vertices and that*

$$\mu = \max\left\{|\lambda| \,\middle|\, \lambda \in \text{Spectrum } A, |\lambda| \neq k\right\}.$$

Then, for all sets S, T of vertices of X, we have

$$\left| E(S,T) - \frac{k\,|S|\,|T|}{n} \right| \leq \mu\sqrt{|S|\,|T|}.$$

Proof By our hypotheses $\mu < k$.

Let δ_S denote the vector whose entries are 1 for the vertices of S and 0 otherwise. Recall the spectral theorem for the symmetric matrix A, the adjacency matrix of X. This says that there is a complete orthonormal basis of $\mathbb{R}^n$ consisting of eigenvectors ϕ_j of A, i.e., for which $A\phi_j = \lambda_j \phi_j$ and

$$(A)_{ab} = \sum_{j=1}^{n} \lambda_j \phi_j(a)\phi_j(b).$$

Here we write $\phi_j(a)$ to denote the entry of ϕ_j corresponding to vertex a of X. We may assume a numbering such that $\phi_1(a) = 1/\sqrt{n}$ and $\lambda_1 = k$. Then, pulling out the first term of the sum gives

$$\begin{aligned} |E(S,T)| = {}^{\mathrm{t}}\delta_S A \delta_T &= \sum_{j=1}^{n} \lambda_j \sum_{\substack{a\in S\\ b\in T}} \phi_j(a)\phi_j(b) \\ &= \frac{k}{n}|S|\,|T| + \sum_{j=2}^{n} \lambda_j \sum_{\substack{a\in S\\ b\in T}} \phi_j(a)\phi_j(b). \end{aligned}$$

Now, by the definition of μ, since our graph is not bipartite there is only one eigenvalue with absolute value equal to k, and

$$\left| \sum_{j=2}^{n} \lambda_j \sum_{\substack{a\in S\\ b\in T}} \phi_j(a)\phi_j(b) \right| \le \mu \sum_{j=2}^{n} \sum_{\substack{a\in S\\ b\in T}} \left|\phi_j(a)\phi_j(b)\right|.$$

To finish the proof, use the Cauchy–Schwarz inequality. Note that the Fourier coefficients of δ_S with respect to the basis ϕ_j are

$$\langle \delta_S, \phi_j \rangle = \sum_{a\in S} \phi_j(a).$$

This implies by Bessel's equality that

$$|S| = \|\delta_S\|_2^2 = \sum_{j=1}^{n} \langle \delta_S, \phi_j \rangle^2.$$

So, the Cauchy–Schwarz inequality says that

$$\left| \sum_{j=2}^{n} \sum_{\substack{a\in S\\ b\in T}} \phi_j(a)\phi_j(b) \right| \le \left| \sum_{j=1}^{n} \sum_{\substack{a\in S\\ b\in T}} \phi_j(a)\phi_j(b) \right| \le \sqrt{|S|\,|T|}.$$

This completes the proof of the lemma. □

Exercise 9.8 Suppose that the graph X represents a gossip network. Explain how you can use Lemma 9.13 to estimate how many people you need to tell to make sure that over one-half the people hear a given rumor after one iteration. What about two iterations?

9.5 Why do Ramanujan graphs have small diameters?

In this section, we present a theorem of Fan Chung [26] which bounds the diameter of a connected k-regular graph in terms of the second largest eigenvalue in absolute value. We assume that the graph is not bipartite to avoid the problem that $-k$ could also be an eigenvalue. From the theorem, we will see that Ramanujan graphs have as small a diameter as possible for sequences of k-regular graphs with numbers of vertices approaching infinity. Thus, the Ramanujan graphs found by Lubotzky, Phillips, and Sarnak [79] were shown to have small diameters.

Definition 9.14 Define the distance $d(x, y)$ between two vertices x, y of a graph X to be the length of a shortest path connecting the vertices. Then the **diameter** of X is

$$\max_{x,y \in V(X)} d(x, y).$$

Theorem 9.15 (Fan Chung [26]) *Suppose that X is a connected non-bipartite k-regular graph with n vertices and that*

$$\mu = \max\left\{|\lambda| \middle| \lambda \in \text{Spectrum } A, \ |\lambda| \neq k\right\}.$$

Then

$$\text{diameter } X \leq 1 + \frac{\log(n-1)}{\log(k/\mu)}.$$

Proof As in the proof of Lemma 9.13, we will use the spectral theorem for the adjacency matrix A of X. This says that there is a complete orthonormal basis of $\mathbb{R}^n$ consisting of eigenvectors ϕ_j of A, so that $A\phi_j = \lambda_j \phi_j$ and

$$(A)_{a,b} = \sum_{j=1}^{n} \lambda_j \phi_j(a)\phi_j(b),$$

where we write $\phi_j(a)$ to denote the entry of ϕ_j corresponding to vertex a of X. Again assume a numbering system such that $\phi_1(a) = 1/\sqrt{n}$ and $\lambda_1 = k$.

Note that, for vertices a, b of X, we have $(A^t)_{a,b} = \#$ {paths of length t connecting a to b}. If d is the diameter of X then $\left(A^d\right)_{a,b} \neq 0$ for some a, b with $d(a, b) = d$. Then $\left(A^{d-1}\right)_{a,b} = 0$, as there is no shorter path connecting a and b. Therefore, if $t = d - 1$,

$$0 = \left(A^t\right)_{ab} = \sum_{j=1}^{t} \lambda_j^t \phi_j(a)\phi_j(b).$$

Now use the Cauchy–Schwarz inequality to see that

$$\begin{aligned}
0 &\geq \frac{k^t}{n} - \mu^t \sum_{j=2}^{n} \left|\phi_j(a)\right| \left|\phi_j(b)\right| \\
&\geq \frac{k^t}{n} - \mu^t \left(\sum_{j=2}^{n} \left|\phi_j(a)\right|^2\right)^{1/2} \left(\sum_{j=2}^{n} \left|\phi_j(b)\right|^2\right)^{1/2} \\
&= \frac{k^t}{n} - \mu^t \sqrt{1 - \phi_1(a)^2}\sqrt{1 - \phi_1(b)^2} = \frac{k^t}{n} - \mu^t \left(1 - \frac{1}{n}\right).
\end{aligned}$$

This implies that

$$\frac{k^t}{n} \leq \mu^t \left(1 - \frac{1}{n}\right).$$

Thus

$$\left(\frac{k}{\mu}\right)^t \leq n - 1.$$

Taking logs,

$$t \log \frac{k}{\mu} \leq \log(n - 1).$$

So, recalling that $t = d - 1$, we have

$$d - 1 \leq \frac{\log(n - 1)}{\log(k/\mu)}.$$

The theorem follows. □

Exercise 9.9 Compute the diameters of your favorite graphs, such as K_4, $K_4 - e$, the Paley graphs, the two-dimensional Euclidean graphs, and the icosahedron.

10

Graph theory prime number theorem

Our main application of the Ihara zeta function is to give an asymptotic estimate for $\pi(m)$, the number of primes of length m in our graph. This is the content of the next theorem. We will use results proved in Chapter 4 on the Ruelle zeta. Before we do this, we need to consider the generating function obtained from the logarithmic derivative of the Ihara zeta function. First recall Definition 4.2 of the numbers N_m and the generating function

$$u\frac{d}{du}\log\zeta_X(u) = \sum_{m\geq 1} N_m u^m. \tag{10.1}$$

This follows from formula (4.5).

Theorem 10.1 (Graph prime number theorem) *We will assume that the graph X satisfies our usual hypotheses (stated in Section 2.1). Suppose that R_X is as in Definition 2.4. If the prime counting function $\pi(m)$ and the g.c.d. of the prime path lengths Δ_X are as in Definitions 2.11 and 2.12 then $\pi(m) = 0$ unless Δ_X divides m. If Δ_X divides m, we have*

$$\pi(m) \sim \Delta_X \frac{R_X^{-m}}{m} \qquad \text{as } m \to \infty.$$

Proof We can imitate the proof of the analogous result for zeta functions of function fields in Rosen [104]. Observe that the defining formula for the Ihara zeta function can be written as

$$\zeta_X(u) = \prod_{n\geq 1}\left(1-u^n\right)^{-\pi(n)}.$$

Then

$$u \frac{d}{du} \log \zeta_X(u) = \sum_{n \geq 1} \frac{n\,\pi(n) u^n}{1 - u^n} = \sum_{m \geq 1} \sum_{d|m} d\,\pi(d) u^m.$$

Here the inner sum is over all positive divisors of m. Thus from formula (10.1) we obtain the **relation between** N_m **and** $\pi(n)$.

$$N_m = \sum_{d|m} d\,\pi(d). \tag{10.2}$$

This sort of relation occurs frequently in number theory and combinatorics. It is inverted using the **Möbius function** $\mu(n)$ defined by

$$\mu(n) = \begin{cases} 1, & n = 1, \\ (-1)^r, & n = p_1 \cdots p_r \text{ for distinct primes } p_i, \\ 0 & \text{otherwise.} \end{cases}$$

Then, by the **Möbius inversion formula**, proved for example in Apostol [3],

$$\pi(m) = \frac{1}{m} \sum_{d|m} \mu\left(\frac{m}{d}\right) N_d. \tag{10.3}$$

Next we look at the two-term determinant formula (4.4), where the **edge adjacency matrix** W_1 is from Definition 4.1. This gives

$$u \frac{d}{du} \log \zeta_X(u) = u \frac{d}{du} \sum_{\lambda \in \operatorname{Spec} W_1} \log(1 - \lambda u) = \sum_{\lambda \in \operatorname{Spec} W_1} \sum_{n \geq 1} (\lambda u)^n.$$

It follows from formula (10.1) that we have a **formula relating** N_m **and the spectrum of** W_1:

$$N_m = \sum_{\lambda \in \operatorname{Spec} W_1} \lambda^m. \tag{10.4}$$

The dominant terms in this last sum are those coming from $\lambda \in \operatorname{Spec} W_1$ such that $|\lambda| = R^{-1}$, with $R = R_X$ from Definition 2.4.

By Theorem 8.1 of Kotani and Sunada [72], the largest absolute value of an eigenvalue λ occurs Δ_X times, and these eigenvalues have the form $e^{2\pi i a/\Delta_X} R^{-1}$ where $a = 1, \ldots, \Delta_X$. Using the orthogonality relations for exponential sums (see Exercise 10.1 and [133]), which are basic to the theory of the finite Fourier transform, we see that

$$\pi(n) \sim \frac{1}{n} \sum_{|\lambda| \text{ maximal}} \lambda^n = \frac{R^{-n}}{n} \sum_{a=1}^{\Delta_X} e^{2\pi i a n/\Delta_X}$$

$$= \frac{R^{-n}}{n} \begin{cases} 0, & \Delta_X \text{ does not divide } n, \\ \Delta_X, & \Delta_X \text{ divides } n. \end{cases} \tag{10.5}$$

The graph prime number theorem follows from formulas (10.3)–(10.5). □

Exercise 10.1 Prove that

$$\sum_{a=1}^{\Delta_X} e^{2\pi i a n/\Delta_X} = \begin{cases} 0, & \Delta_X \text{ does not divide } n, \\ \Delta_X, & \Delta_X \text{ divides } n. \end{cases}$$

Example 10.2: Primes in $K_4 - e$, the graph obtained from K_4 by deleting an edge e See Figure 2.5. We have seen that

$$\zeta_X(u)^{-1} = (1-u^2)(1-u)(1+u^2)(1+u+2u^2)(1-u^2-2u^3).$$

From this, we have

$$u\frac{d}{du}\log\zeta_X(u) = 12u^3 + 8u^4 + 24u^6 + 28u^7 + 8u^8 + 48u^9$$
$$+ 120u^{10} + 44u^{11} + 104u^{12} + 416u^{13} + 280u^{14} + O(u^{15}).$$

Now we want to use $N_m = \sum_{d|m} d\,\pi(d)$ to compute the small values of $\pi(m)$. First, $3\pi(3) = 12$ implies that $\pi(3) = 4$ and we can find $\pi(4)$ and $\pi(5)$ similarly:

$$\pi(3) = 4, \qquad \pi(4) = 2, \qquad \pi(5) = 0.$$

Then $6\pi(6) + 3\pi(3) = 24$ implies that $\pi(6) = 2$. Next we find that $\pi(7) = 4$, while $\pi(8) = 0$. Then $9\pi(9) + 3\pi(3) = 48$ implies that $\pi(9) = 36/9 = 4$. Finally $10\pi(10) + 5\pi(5) = 120$ says that $\pi(10) = 12$.

So the rest of our list reads

$$\pi(6) = 2, \qquad \pi(7) = 4, \qquad \pi(8) = 0, \qquad \pi(9) = 4, \qquad \pi(10) = 12.$$

The reader should look at the graph for examples of primes of lengths $3, 4, 6, 7, 9, 10$.

Exercise 10.2 If the graph $X = K_4$, the tetrahedron, find $\pi(m)$ for $m = 3, 4, 5, \ldots, 11$.

If the Riemann hypothesis (either version for irregular graphs) holds for $\zeta_X(u)$ then one has a good bound on the error term in the prime number theorem, by formulas (10.3) and (10.4). By Theorem 9.8 (the Alon–Boppana theorem), the bound on the error term will be best possible for a family of connected $(q + 1)$-regular graphs whose number of vertices approaches infinity.

10.1 Which graph properties are determined by the Ihara zeta?

There are many papers on this question. See Yaim Cooper [24] Debra Czarneski [33], Matthew Horton [57], [58], and Christopher Storm [125], for example. Some things are obvious. The degree of the reciprocal of the Ihara zeta is the number of directed edges. The number of vertices is found by noting that the rank r of the fundamental group is determined as in the next paragraph and that $r - 1 = |E| - |V|$. We have also seen that the zeta function tells us the numbers N_m and thus the **girth** (the length of the shortest cycle) of the graph, which is the first m with non-zero N_m.

Horton [57], [58] found a simple formula for the girth of the graph from the reciprocal of zeta. He also showed that the chromatic polynomial of the graph (discussed in Bollobás [19] for example) cannot be determined from zeta alone. Storm [125] found that zeta determines the clique number as well as the number of Hamiltonian cycles. A **clique** is a complete graph that is an induced subgraph. A **Hamiltonian cycle** is a cycle that visits every vertex exactly once. Later (see Chapter 21), we will find that it is possible for $\zeta_X(u) = \zeta_Y(u)$ to hold even if graph X is not isomorphic to graph Y. Thus zeta does **not** determine the graph up to graph isomorphism.

The Ihara zeta function determines the rank of the fundamental group, for the latter is the order of the pole of the Ihara zeta function at $u = 1$. The **complexity** κ_X of a graph is defined to be the number of spanning trees in X. One can use the matrix-tree theorem (see Biggs [15]) to prove that

$$\left. \frac{d^r}{du^r} \zeta_X^{-1}(u) \right|_{u=1} = r!(-1)^{r+1} 2^r (r - 1)\kappa_X. \tag{10.6}$$

This result is an exercise on the last page of Terras [133], where some hints are given. It is an analog of the formula for the Dedekind zeta function of a number field at 0 (a formula involving the class number and the regulator of the number field). See Figure 1.3 and Lang [73].

Exercise 10.3 Prove formula (10.6).

Exercise 10.4 Prove the prime number theorem for a $(q + 1)$-regular graph using Theorem 2.5, Ihara's theorem, with the three-term determinant rather than the $\det(I - W_1)^{-1}$ formula.

We have found that the Ihara zeta function possesses many analogous properties to the Dedekind zeta function of an algebraic number field. There are other analogs as well. For example there is an analog of the ideal class group known as the Jacobian of a graph. It has order equal to κ_X, the complexity. It has been considered by Bacher, de la Harpe and Nagnibeda [5] as well as Baker and Norine [7].

Part III
Edge and path zeta functions

In Part III we consider two multivariable zeta functions associated with a finite graph, the edge zeta and the path zeta. We will give a matrix analysis version of Bass's proof of Ihara's determinant formula. This implies that there is a determinant formula for the vertex zeta function of weighted graphs even if the weights are not integers. We will discuss what deleting an edge of a graph (**fission**) does to the edge zeta function. We will also discuss what happens if a graph edge is **fused**, i.e., shrunk to a point. There is an application of the edge zeta to error correcting codes, which will be discussed in the last part of this book. See also Koetter *et al.* [70] and [71].

11

Edge zeta functions

11.1 Definitions and Bass's proof of the Ihara three-term determinant formula

Notation 11.1 From now on, we will change our notation for the Ihara zeta function of earlier chapters, replacing $\zeta_X(u)$ by $\zeta(u, X)$ (or even $\zeta_V(u, X)$, where the subscript V is for **vertex**). We may call the Ihara zeta a "**vertex zeta**," although we will try to avoid this.

Definition 11.2 The **edge matrix** W for graph X is a $2m \times 2m$ matrix with a, b entry corresponding to the oriented edges a and b. This a, b entry is a complex variable w_{ab} if edge a feeds into edge b and $b \neq a^{-1}$ and is 0 otherwise.

Note that W_1 from Definition 4.1 is obtained from the edge matrix W by setting all non-zero entries of W equal to 1.

Definition 11.3 Given a closed path C in X, which is written as a product of oriented edges $C = a_1 a_2 \cdots a_s$, the **edge norm** of C is

$$N_E(C) = w_{a_1 a_2} w_{a_2 a_3} \cdots w_{a_{s-1} a_s} w_{a_s a_1}.$$

The **edge zeta function** is

$$\zeta_E(W, X) = \prod_{[P]} (1 - N_E(P))^{-1},$$

where the product is over primes in X. Here we will assume that all $|w_{ab}|$ are sufficiently small for convergence.

Specializing variables to obtain other zetas involves the following.

(1) Clearly if you set all non-zero variables in W equal to $u \in \mathbb{C}$, the edge norm $N_E(C)$ specializes to $u^{\nu(C)}$. Therefore (by the definitions of the Ihara zeta function and the edge zeta function) the **edge zeta function specializes to the Ihara (vertex) zeta function**; i.e.,

$$\zeta_E(W, X)\big|_{0 \neq w_{ab} = u} = \zeta(u, X). \tag{11.1}$$

(2) If X is a **weighted graph** with weight function L, and you specialize the non-zero variables

$$w_{ab} = u^{(L(a)+L(b))/2}, \tag{11.2}$$

you get the weighted Ihara zeta function of Definition 6.2. Or you could specialize by setting

$$w_{ab} = u^{L(a)}. \tag{11.3}$$

As in Mizuno and Sato [89], one can also associate non-negative values w_e with each directed edge e and then write $w_{ab} = \sqrt{w_a}\sqrt{w_b}$, if directed edge a leads into directed edge b without backtracking, and $w_{ab} = 0$ otherwise. This leads to a nice version of the zeta for weighted graphs.

(3) To obtain the Hashimoto edge zeta function discussed in Stark and Terras [119], specialize by setting $w_{ab} = u_a$. This is the zeta in the application to error-correcting codes in Chapter 24.

(4) If you cut or delete an edge of a graph (which we view as "**fission**"), you can compute the edge zeta for the new graph with one less edge by setting any variable equal to 0 if the cut or deleted edge or its inverse appear in one of its subscripts. Note that graph theorists usually call an edge a "cut edge" only if its removal disconnects the graph.

(5) You can also use the variables w_{ab} in the edge matrix W corresponding to $b = a^{-1}$ to produce a zeta function that keeps track of paths with backtracking or tails. See Bartholdi [11].

(6) Finally, one can consider edge zetas of directed graphs. See Horton [57], [58].

The edge zeta again has a determinant formula and is the reciprocal of a polynomial in the w_{ab} variables. This is stated in the following theorem, whose proof should be compared with that of formula (4.4).

Theorem 11.4 (Determinant formula for the edge zeta) *we have*

$$\zeta_E(W, X) = \det(I - W)^{-1}.$$

Proof First note that, from the Euler product for the edge zeta function, we have

$$-\log \zeta_E(W, X) = \sum_{[P]} \sum_{j \geq 1} \frac{1}{j} N_E(P)^j.$$

Since there are $\nu(P)$ elements in the prime $[P]$, we have

$$-\log \zeta_E(W, X) = \sum_{\substack{m \geq 1 \\ j \geq 1}} \frac{1}{jm} \sum_{\substack{P \\ \nu(P)=m}} N_E(P)^j.$$

Here the sum is over primitive cycles P.

It follows that

$$-\log \zeta_E(W, X) = \sum_C \frac{1}{\nu(C)} N_E(C).$$

Here we sum over paths C which need not be prime paths but are still closed and without backtracking or tails. This comes from the fact that such a path C has the form P^j, for some prime path P and $j = 1, 2, \ldots$ Then, by the exercise below, we see that

$$-\log \zeta_E(W, X) = \sum_{m \geq 1} \frac{1}{m} \operatorname{Tr} W^m.$$

Finally, again from the exercise below, we see that the right-hand side of the preceding formula is $\log \det(I - W)^{-1}$. This proves the logarithm of the determinant formula, and taking the exponential of both sides gives the theorem. □

Exercise 11.1 Prove that

$$\sum_C \frac{1}{\nu(C)} N_E(C) = \sum_{m \geq 1} \frac{1}{m} \operatorname{Tr} W^m = -\operatorname{Tr} \log(I - W) = \log \det(I - W)^{-1}.$$

Hints:

(1) For the first equality, you need to think about $\operatorname{Tr} W^m$ as an $(m+1)$-fold sum of products of the w_{ij} and relate this to closed paths C of length m.
(2) For the second equality, use the power series for $\log(I - W)$.
(3) Recall Exercise 4.1.

By formula (11.1) we have the following corollary, since specializing all the non-zero variables in W to u yields the matrix uW_1, where W_1 is from

Definition 4.1. We also proved this corollary in Chapter 4 on Ruelle zeta functions.

Corollary 11.5 *We have* $\zeta_X(u) = \zeta_V(u, X) = \det(I - uW_1)^{-1}$.

Moral The poles of $\zeta_X(u)$ are the reciprocals of the eigenvalues of W_1.

Exercise 11.2 Write W_1 in block form with $|E| \times |E|$ blocks:

$$W_1 = \begin{pmatrix} A & B \\ C & D \end{pmatrix}.$$

(a) Show that $D = {}^tA,\ B = {}^tB,\ C = {}^tC$. The diagonal entries of B and C are zero.
(b) Show that the sum of the entries of the ith row of W_1 is the degree of the vertex which is the starting vertex of edge i.

Exercise 11.3 Consider some weighted graphs and their zeta functions. Can you expect them to have all the properties of the ordinary Ihara zetas, no matter what sort of weights are involved?

Example 11.6: Dumbbell graph Figure 11.1 shows a dumbbell graph X. For this graph we find that

$$\zeta_E(W, X)^{-1} = \det \begin{pmatrix} w_{11}-1 & w_{12} & 0 & 0 & 0 & 0 \\ 0 & -1 & w_{23} & 0 & 0 & w_{26} \\ 0 & 0 & w_{33}-1 & 0 & w_{35} & 0 \\ 0 & w_{42} & 0 & w_{44}-1 & 0 & 0 \\ w_{51} & 0 & 0 & w_{54} & -1 & 0 \\ 0 & 0 & 0 & 0 & w_{65} & w_{66}-1 \end{pmatrix}.$$

Note that if we cut or delete the vertical edges, i.e., edges e_2 and e_5, we should specialize all the variables with 2 or 5 in them to 0. This yields the edge zeta function of the subgraph with the vertical edge removed and incidentally diagonalizes the matrix W. We call this "**fission**." The edge zeta is particularly suited to keeping track of such fission.

Exercise 11.4 Do another example in which you compute the edge zeta function of your favorite graph. Then see what happens if you delete an edge.

Next we give a version of **Bass's proof of the Ihara determinant formula** (Theorem 2.5) using the preceding theorem. In what follows, n is the number of vertices of X and m is the number of unoriented edges of X.

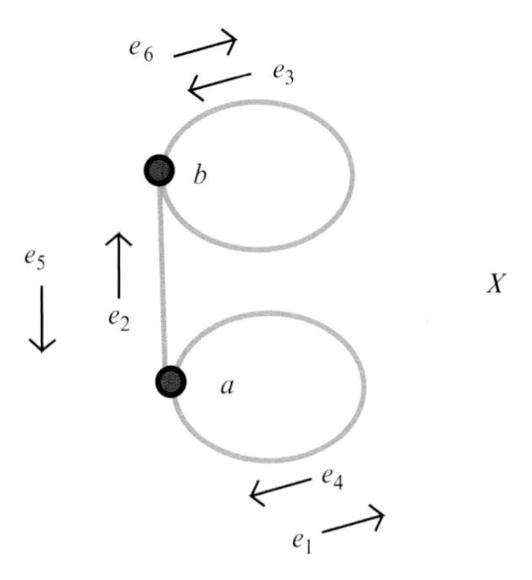

Figure 11.1 The dumbbell graph.

First define some matrices. Set

$$J = \begin{pmatrix} 0 & I_m \\ I_m & 0 \end{pmatrix}.$$

Then define the $n \times 2m$ **start matrix** S and the $n \times 2m$ **terminal matrix** T by setting

$$s_{ve} = \begin{cases} 1 & \text{if } v \text{ is the starting vertex of edge } e, \\ 0 & \text{otherwise} \end{cases}$$

and

$$t_{ve} = \begin{cases} 1 & \text{if } v \text{ is the terminal vertex of edge } e, \\ 0 & \text{otherwise.} \end{cases}$$

Note that $S = (MN)$, $T = (NM)$, where M and N are $|V| \times |E|$ matrices, thanks to our numbering system for the directed edges, in which $e_{j+|\mathrm{E}|} = e_j^{-1}$. Here $j = 1, 2, \ldots, |E|$.

Exercise 11.5 Write $S = (MN)$, $T = (NM)$, where M and N are $|V| \times |E|$ matrices of 0's and 1's. Use the following proposition to create random graphs and plot the poles of their zeta functions.

Hint: Make use of the Matlab commands creating random permutation matrices P_i to build up $M = (P_1 \cdots P_k)$. Build up N similarly. Then obtain W_1 from Proposition 11.7 below. We created Figure 11.2 below this way.

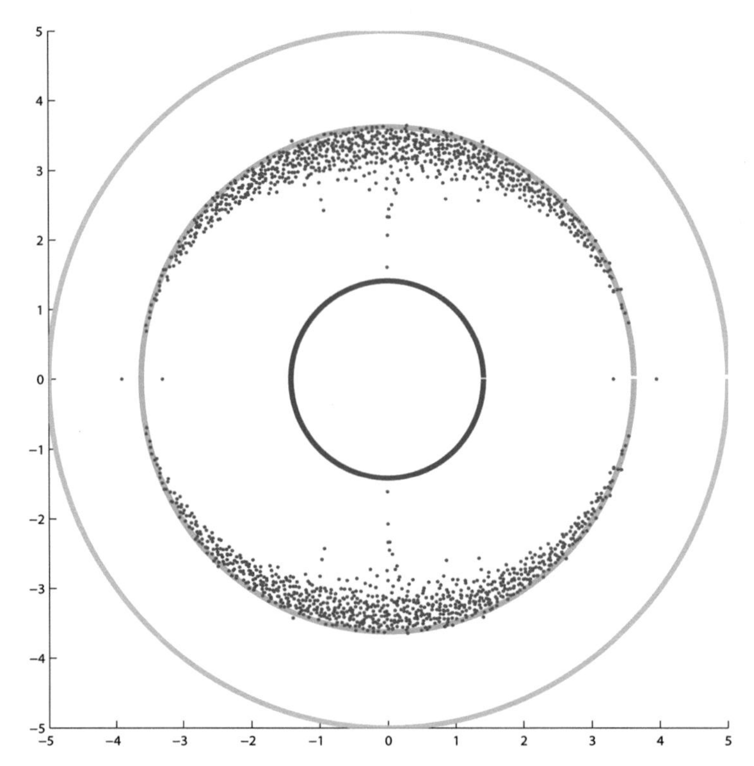

Figure 11.2 A Matlab experiment. The eigenvalues ($\neq \pm 1$ or $1/R$) of the edge adjacency matrix W_1 for a random graph are the purple points. The innermost circle has radius $\sqrt{p}$ and the middle circle, shown in green, has radius $1/\sqrt{R}$. The outermost circle has radius $\sqrt{q}$. The green circle is the RH circle. Because the eigenvalues of W_1 are reciprocals of the poles of zeta, now the RH says the spectrum should be inside the green circle. The RH appears to be approximately true. The graph has 800 vertices, mean degree $\cong 13.125$, and edge probability $\cong 0.0164$.

Proposition 11.7 (Some matrix identities) *Using the preceding definitions, the following formulas hold. (As before, we write ${}^{t}M$ for the transpose of the matrix M).*

(1) $SJ = T$ and $TJ = S$.
(2) If A is the adjacency matrix of X and $Q + I_n$ is the diagonal matrix whose jth diagonal entry is the degree of the jth vertex of X then $A = S\,{}^{t}T$ and $Q + I_n = S\,{}^{t}S = T\,{}^{t}T$.
(3) The edge adjacency matrix W_1 from Definition 4.1 satisfies $W_1 + J = {}^{t}TS$.

Proof Part (1) This comes from the fact that the starting (terminal) vertex of edge e_j is the terminal (starting) vertex of edge $e_{j+|E|}$, according to our edge numbering system from formula (2.1).

Part (2) Consider

$$(S\,{}^{\mathrm{t}}T)_{ab} = \sum_{e} s_{ae} t_{be}.$$

The right-hand side is the number of oriented edges e such that a is the initial vertex and b the terminal vertex of e, which is the ab entry of A. Note that A_{aa} is twice the number of loops at vertex a. Similar arguments prove the second formula.

Part (3) We have

$$({}^{\mathrm{t}}TS)_{ef} = \sum_{v} t_{ve} s_{vf}.$$

The sum is 1 iff edge e feeds into edge f, even if $f = e^{-1}$. So, recalling our directed edge labeling convention, if $f \neq e^{-1}$ then we get $(W_1)_{ef} = (W_1 + J)_{ef}$ but if $f = e^{-1}$ then we get $(J)_{ef} = (W_1 + J)_{ef}$. □

Exercise 11.6 (a) Prove part (1) of Proposition 11.7.
(b) Prove that $Q + I_n = S\,{}^{\mathrm{t}}S$.

Finally we come to the proof that we have been advertising, **Bass's proof of the Ihara determinant formula, Theorem 2.5.**

Proof of Theorem 2.5 We seek to derive the Ihara determinant formula

$$\zeta_X(u)^{-1} = \left(1 - u^2\right)^{r-1} \det\left(I - Au + Qu^2\right)$$

from the identity $\zeta_X(u)^{-1} = \det(I - W_1 u)$, which was Corollary 11.5 above. This will be done using some simple block-matrix identities.

In the following identity all matrices are $(n + 2m) \times (n + 2m)$, where the first block is $n \times n$, n being the number of vertices of X and m the number of unoriented edges of X. Use the Proposition 11.7 to see that

$$\begin{pmatrix} I_n & 0 \\ {}^{\mathrm{t}}T & I_{2m} \end{pmatrix} \begin{pmatrix} I_n(1 - u^2) & Su \\ 0 & I_{2m} - W_1 u \end{pmatrix}$$

$$= \begin{pmatrix} I_n - Au + Qu^2 & Su \\ 0 & I_{2m} + Ju \end{pmatrix} \begin{pmatrix} I_n & 0 \\ {}^{\mathrm{t}}T - {}^{\mathrm{t}}Su & I_{2m} \end{pmatrix}.$$

□

Exercise 11.7 Check this equality. Relate it to the Schur complement of a block in a matrix.

Now take determinants to obtain

$$(1-u^2)^n \det\big(I - W_1 u\big) = \det\big(I_n - Au + Qu^2\big)\det\big(I_{2m} + Ju\big).$$

To finish the proof of Theorem 2.5, observe that

$$I + Ju = \begin{pmatrix} I & Iu \\ Iu & I \end{pmatrix}$$

implies that

$$\begin{pmatrix} I & 0 \\ -Iu & I \end{pmatrix}(I + Ju) = \begin{pmatrix} I & Iu \\ 0 & I(1-u^2) \end{pmatrix}.$$

Thus $\det(I + Ju) = (1-u^2)^m$. Since $r - 1 = m - n$ for a connected graph, Theorem 2.5 follows. □

11.2 Properties of W_1 and a proof of the theorem of Kotani and Sunada

Next we want to prove Theorem 8.1, due to Kotani and Sunada. First we will need some facts from linear algebra as well as some facts about the W_1 matrix. In the next definition, a **permutation matrix** is a square matrix such that exactly one entry in each row and each column is 1 and the rest of the entries are 0.

Definition 11.8 An $s \times s$ matrix A, with $s > 1$, whose entries are non-negative is **irreducible** iff there does **not** exist a permutation matrix P such that

$$A = {}^{t}P \begin{pmatrix} B & C \\ 0 & D \end{pmatrix} P,$$

where B is a $t \times t$ matrix with $1 \le t < s$.

The following theorem is proved in Horn and Johnson [56], p. 361. The Perron–Frobenius theorem concerns such matrices.

Theorem 11.9 *An $n \times n$ matrix A with all non-negative entries is irreducible iff $(I + A)^{n-1}$ has all positive entries.*

Theorem 11.10 (Facts about W_1) *Assume that X satisfies the usual hypotheses, stated in Section 2.1. Let $n = |V|$ be the number of vertices of graph X and let $m = |E|$ be the number of undirected edges of X.*

(1) The jth row sum of the entries of W_1 is $q_j = -1$ plus the degree of the starting vertex of the jth edge.
(2) (Horton) The singular values of W_1 (i.e., the square roots of the eigenvalues of $W_1\,{}^{t}W_1$) are $\{q_1, \ldots, q_n, \underbrace{1, \ldots, 1}_{2m-n}\}$.
(3) The matrix $(I + W_1)^{2m-1}$ has all positive entries. This means that the matrix W_1 is irreducible.

Proof Part (1) We leave this as an **exercise**.

Part (2) (Horton [57], [58]). Modify W_1 to list together all edges ending at the same vertex. Note that then

$$W_1\,{}^{t}W_1 = \begin{pmatrix} A_1 & 0 & 0 \\ \vdots & \ddots & \vdots \\ 0 & 0 & A_n \end{pmatrix} \qquad \text{and} \qquad A_j = (q_j - 1)J + I,$$

where J is a $(q_j + 1) \times (q_j + 1)$ matrix of 1's. Since the spectrum of J is $\{q_j + 1, 0, \ldots, 0\}$, the spectrum of A_j is $\{q_j^2, 1, \ldots, 1\}$. The result follows.

(3) This follows from Lemma 11.11 below. □

Exercise 11.8 Consider the graph which consists of one vertex with two loops and another vertex on one of these loops. Modify W_1 to list all edges ending at the same vertex together and compute $W_1\,{}^{t}W_1$.

Lemma 11.11 *Suppose that X satisfies our usual hypotheses, stated in Section 2.1. Given a directed edge e_1 in X starting at a vertex v_1 and a directed edge e_2 in X ending at a vertex v_2 ($v_1 = v_2$, $e_1 = e_2$, and $e_1 = e_2^{-1}$ are allowed), there exists a backtrackless path $P = P(e_1, e_2)$ in X from v_1 to v_2 with initial edge e_1, terminal edge e_2, and length $\leq 2|E|$.*

Proof See Figure 11.3, which shows our construction of $P(e_1, e_2)$ for two alternative cases. First we construct a path P without worrying about its length. This construction is not minimal, but there are relatively few possibilities to consider.

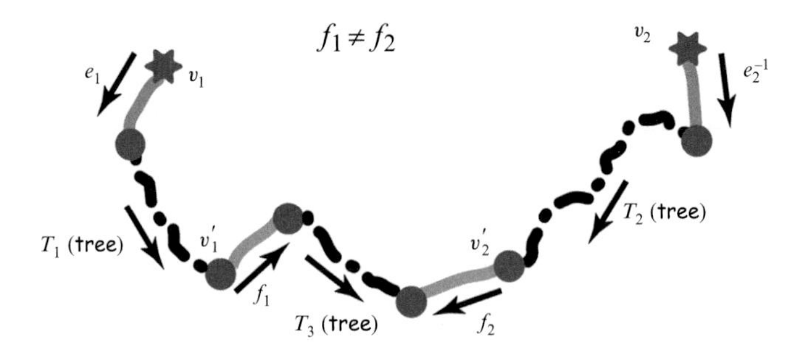

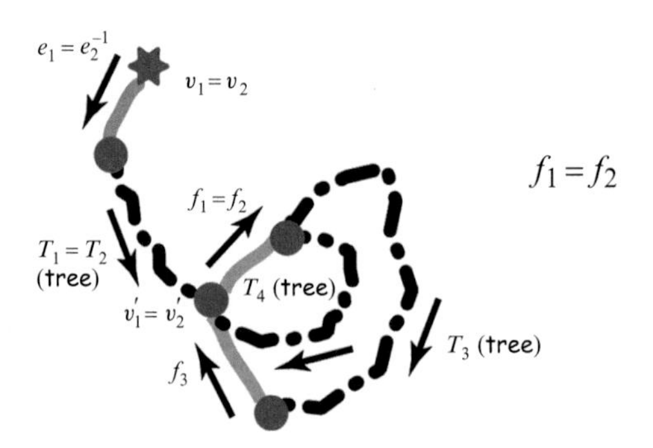

Figure 11.3 The paths in Lemma 11.11. Here the broken-and-dotted paths are along the spanning tree of X. The edges e_1 and e_2 may not be the edges of X which are cut to get the spanning tree T. But f_1, f_2, and (in the lower diagram) f_3 are cut or non-tree edges. The lower diagram does not show the most general situation, as f_3 need not touch $f_1 = f_2$.

Choose a spanning tree T of X. Define a **cut** edge of X to mean an edge left out of T. Begin by creating two backtrackless paths $P_1 f_1$ and $P_2 f_2$ with initial edges e_1 and e_2^{-1} and terminal edges f_1 and f_2 such that f_1 and f_2 are cut edges (i.e., non-tree edges of X). If e_1 is a cut edge, let P_1 have length 0 and let $f_1 = e_1$ (i.e., $P_1 f_1 = e_1$). If e_1 is not a cut edge, take P_1 to be a backtrackless path in T with initial edge e_1 which proceeds along T until it is impossible to go any further along the tree. Symbolically we write $P_1 = e_1 T_1$, where T_1 is a path along the tree, possibly of length 0.

Let v_1' be the terminal vertex of P_1. With respect to the tree T, v_1' is a dangler or leaf (a vertex of degree 1), but X has no danglers. Thus there must be a directed cut edge in X, which we take to be f_1, with initial vertex v_1'. By construction $P_1 f_1$ is backtrackless also since P_1 is in the tree and f_1 is not.

Similarly, if e_2 is a cut edge then let P_2 have length 0 and let $f_2 = e_2^{-1}$ (i.e., $P_2 f_2 = e_2^{-1}$). If e_2 is not a cut edge then as above form a backtrackless path $P_2 f_2 = e_2^{-1} T_2 f_2$, where T_2 is in the tree and is possibly of length 0 and f_2 is a cut edge. In all cases we let v_1' and v_2' be the initial vertices of f_1 and f_2.

Now, if we can find a path P_3 beginning at the terminal vertex of f_1 and ending at the terminal vertex of f_2 such that the path $f_1 P_3 f_2^{-1}$ has no backtracking then $P = P_1 f_1 P_3 f_2^{-1} P_2^{-1}$ will have no backtracking, with e_1 and e_2 as its initial and terminal edges, respectively. Of course, creating the path $f_1 P_3 f_2^{-1}$ was the original problem in proving the lemma. However, we now have the additional information that f_1 and f_2 are cut edges of the graph X.

We now have two cases. In case 1 we have $f_1 \neq f_2$, shown in the upper diagram in Figure 11.3. For this case we can take $P_3 = T_3$, the path within the tree T running from the terminal vertex of f_1 to the terminal vertex of f_2. Then, even if the length of T_3 is 0, the path $f_1 T_3 f_2^{-1}$ has no backtracks and we have created P.

In case 2 we have $f_1 = f_2$. Thus f_1 and f_2 are the same cut edge of X. In the worst-case scenario, we would have $e_2 = e_1^{-1}$, $T_2 = T_1$, and $f_2 = f_1$. See the lower diagram in Figure 11.3. Since the usual hypotheses say that X has rank at least 2, there must be another cut edge f_3 of X with $f_3 \neq f_1$ or f_1^{-1}. Let T_3 be the path along the tree T from the terminal vertex of f_1 to the initial vertex of f_3 and let T_4 be the path along the tree T from the terminal vertex of $f_2 = f_1$ to the terminal vertex of f_3. Then $P_3 = T_3 f_3 T_4^{-1}$ has the desired property that $f_1 P_3 f_1^{-1}$ has no backtracking, even if T_3 and/or T_4 have length 0. Thus we have created in all cases a backtrackless path P with initial edge e_1 and terminal edge e_2.

Now one can create a path P of length $\leq 2|E|$ as follows. If an edge is repeated, it is possible to delete all the edges in between the first and second versions of that edge as well as the 2nd version itself, without harming the properties of P. □

Exercise 11.9 Prove Lemma 11.11 in the case where the graph X is a bouquet of n loops with $n \geq 2$.

Corollary 11.12 *Suppose that X satisfies our usual hypotheses; in particular suppose that X is finite and connected with no degree-1 vertices and that its fundamental group has rank at least 2. Then the edge adjacency matrix W_1 is irreducible.*

Proof It follows from the preceding lemma that all entries of $(I + W_1)^{2|E|-1}$ are positive. To see this, we look at the ef entry of $W_1^{\nu-1}$. Take a backtrackless path P starting at e and ending at f. The lemma says that we can assume the length of P is $\nu = \nu(P) \leq 2|E|$. The e, f entry of the matrix $W_1^{\nu-1}$ is a sum of terms of the form $w_{e_1 e_2} \cdots w_{e_{\nu-1} e_\nu}$, where each e_i denotes an oriented edge and $e_1 = e$, $e_\nu = f$. The term corresponding to the path P will be positive and the rest of the terms are non-negative. Then use Theorem 11.9. □

Example 11.13: The shortest path from Lemma 11.11 for a dumbbell Consider the dumbbell graph in Figure 11.1. The shortest possible path $P(e_1, e_1^{-1})$, using the terminology of Lemma 11.11, is $e_1 e_2 e_3 e_2^{-1} e_1^{-1}$ with length $5 = 2|E| - 1$.

Exercise 11.10 Show that the hypotheses on X in Lemma 11.11 are necessary. In particular, what happens when the rank of the fundamental group is 1? And what happens if some vertices have degree 1?

We have the following corollary to Theorem 11.10.

Corollary 11.14 *The poles of the Ihara zeta function of X are contained in the region* $1/q \leq R_X \leq |u| \leq 1$, *where* $q+1$ *is the maximum degree of a vertex of X.*

Proof The poles are reciprocals of the eigenvalues of the W_1 matrix. The singular values of W_1 are

$$\Big\{q_1, \ldots, q_n, \underbrace{1, \ldots, 1}_{2m-n}\Big\}.$$

Assume that $q_j \geq q_{j+1}$. This means that q_j^2 and 1 are successive maxima of the Rayleigh quotient ${}^t\overline{(W_1 v)} W_1 v / ({}^t\overline{v} v)$ over vectors $v \in \mathbb{C}^n$ that are orthogonal to the vectors at which the preceding maxima are taken. If $W_1 v = \lambda v$ for non-zero v then the Rayleigh quotient is $|\lambda|^2$. □

Next recall the Perron–Frobenius theorem. A proof can be found in Horn and Johnson [56]. We first recall a definition.

Definition 11.15 The **spectral radius** $\rho = \rho(A)$ of a matrix A is defined to be the maximum of all $|\lambda|$, for λ an eigenvalue of A.

Theorem 11.16 (Perron and Frobenius) *Let A be an* $s \times s$ *matrix all of whose entries are non-negative. Assume that A is irreducible. Then we have the following facts.*

(1) *The spectral radius* $\rho(A)$ *is positive and is an eigenvalue of* A *which is simple (both in the algebraic and geometric senses). There is a corresponding eigenvector of* A *all of whose entries are positive.*
(2) *Let* $S = \{\lambda_1, \ldots, \lambda_k\}$ *be the eigenvalues of* A *having maximum modulus. Then*

$$S = \left\{ \rho(A) e^{2\pi i a/k} \middle| \, a = 1, \ldots, k \right\}.$$

(3) *The spectrum of* A *is invariant under rotation by* $2\pi/k$.

If $A = W_1$, the spectral radius $\rho(W_1) = R^{-1}$ and the number k is Δ, the g.c.d. of the lengths of the primes of X.

Exercise 11.11 Prove the last statement.
Hint: Part (3) of Theorem 11.16 implies that $\zeta_X(u)^{-1} = \det(I - W_1 u) = f(u^k)$. By the definition of Δ, we have $\log \zeta_X(u)^{-1} = F(u^\Delta)$. Thus $\pi(m) = 0$ unless Δ divides m. Recall that if $\zeta = e^{2\pi i/k}$ then

$$\prod_{j=1}^{k} \left(1 - \zeta^j u\right) = 1 - u^k.$$

Now we proceed to prove the **Kotani and Sunada theorem**, Theorem 8.1. Let us restate what we are proving. Suppose that $q + 1$ **is the maximum degree and** $p + 1$ **is the minimum degree of a graph** X satisfying our usual hypotheses, stated in Section 2.1.

(1) **Every pole** u **of** $\zeta_X(u)$ **satisfies** $R_X \leq |u| \leq 1$, **with** R_X **from Definition 2.4 and** $q^{-1} \leq R_X \leq p^{-1}$.
(2) **For a graph** X, **every non-real pole** u **of** $\zeta_X(u)$ **satisfies the inequality** $q^{-1/2} \leq |u| \leq p^{-1/2}$.
(3) **The poles of** ζ_X **on the circle** $|u| = R_X$ **have the form** $R_X e^{2\pi i a/\Delta_X}$, **where** $a = 1, \ldots, \Delta_X$. **Here** Δ_X **is from Definition 2.12.**

Proof of Theorem 8.1 The second inequality in part (1) comes from a result of Frobenius stating that $\rho(W_1) = R^{-1}$ is bounded above and below by the maximum and minimum row sums of W_1, respectively. See Minc [88], p. 24, or Horn and Johnson [56], p. 492.

We know that $R_X \leq |u| \leq 1$ by Corollary 11.14. Here we will give the Kotani and Sunada proof that $|u| \leq 1$.

If u is a pole of $\zeta_X(u)$ with $|u| \neq 1$ then there is a non-zero vector f such that $\left(I - Au + u^2 Q\right) f = 0$.

We denote the inner product $\langle f, g \rangle = {}^t\bar{g} f$ for column vectors f, g in $\mathbb{C}^n$. Then

$$0 = \left\langle \left(I - uA + u^2 Q \right) f, f \right\rangle = \|f\|^2 - u\langle Af, f\rangle + u^2 \langle Qf, f\rangle.$$

Set

$$\lambda = \frac{\langle Af, f\rangle}{\|f\|^2}, \qquad \delta = \frac{\langle Df, f\rangle}{\|f\|^2}, \qquad \text{and} \qquad D = Q + I.$$

So we have $1 - u\lambda + u^2(\delta - 1) = 0$. The quadratic formula gives

$$u = \frac{\lambda \pm \sqrt{\lambda^2 - 4(\delta - 1)}}{2(\delta - 1)}.$$

Clearly $p \leq \delta - 1 \leq q$. We also have $|\lambda| \leq \delta$. To prove this, we can make use of the S and T matrices in Proposition 11.7. Note that S and T are formed from the matrices M and N, i.e., $S = (M \; N)$ and $T = (N \; M)$, where M and N have $m = |E|$ columns. Then it is an **exercise** using the matrix identities in Proposition 11.7 to show that

$$D - A = (M - N)\,{}^{t}(M - N) \qquad \text{and} \qquad D + A = (M + N)\,{}^{t}(M + N).$$

So $|\langle Af, f\rangle| \leq \langle Df, f\rangle$. It follows that $|\lambda| \leq \delta$. There are now two cases.

Case 1 The pole u is real.

Then

$$\frac{\lambda + \sqrt{\lambda^2 - 4(\delta - 1)}}{2(\delta - 1)} \leq \frac{\delta + \sqrt{\delta^2 - 4(\delta - 1)}}{2(\delta - 1)} = 1$$

and

$$\frac{\lambda - \sqrt{\lambda^2 - 4(\delta - 1)}}{2(\delta - 1)} \geq \frac{-\delta - \sqrt{\delta^2 - 4(\delta - 1)}}{2(\delta - 1)} = -1.$$

Thus $|u| \leq 1$.

Case 2 The pole u is not real.

Then

$$|u|^2 = \frac{\lambda^2 + \left(4(\delta - 1) - \lambda^2\right)}{4(\delta - 1)^2} = \frac{1}{\delta - 1}.$$

Part (2) of Theorem 8.1 follows from this and the fact that $p \leq \delta - 1 \leq q$.

Finally, part (3) of Theorem 8.1 follows from the Perron–Frobenius Theorem 11.16. □

Proposition 11.17 *Make the usual hypotheses on the graph X. Recall Definitions 2.4 and 11.15. Then we have the inequality*

$$\rho(A) \geq \frac{p}{q} + \frac{1}{R},$$

where $p+1$ is the minimum degree and $q+1$ is the maximum degree among the vertices of X.

Proof Let $u = R$, the radius of convergence of the Ihara zeta or the reciprocal of the Perron–Frobenius eigenvalue of W_1. Then, as in the preceding proof, we have

$$0 = \|f\|^2 - u\langle Af, f\rangle + u^2\langle Qf, f\rangle.$$

Set

$$\lambda = \frac{\langle Af, f\rangle}{\|f\|^2}, \qquad \delta = \frac{\langle Df, f\rangle}{\|f\|^2}, \qquad \text{and} \qquad D = Q + I.$$

So we have $1 - R\lambda + R^2(\delta - 1) = 0$. Thus $\lambda = 1/R + R(\delta - 1)$. It follows that $\rho \geq 1/R + p/q$, since

$$R \geq \frac{1}{q} \qquad \text{and} \qquad \delta - 1 \geq p.$$

□

Problem 11.18 Whether it is possible to improve the inequality in this proposition by replacing p/q with 1 is a research problem.

12
Path zeta functions

Here we look at a zeta function invented by Stark. It has several advantages over the edge zeta. It can be used to compute the edge zeta using smaller determinants. It gives the edge zeta for a graph in which an edge has been fused, i.e., shrunk to one vertex.

First recall that the fundamental group of X can be identified with the group generated by the edges left out of a spanning tree T of X. Then T has $|V| - 1 = n - 1$ edges. We label the oriented versions of these **edges left out of the spanning tree** T ("**cut**" or "**deleted**"edges of T) and their inverses

$$e_1, \quad \ldots, \quad e_r, \quad e_1^{-1}, \quad \ldots, \quad e_r^{-1}.$$

Label the remaining (oriented) **edges in the spanning tree** T

$$t_1, \quad \ldots, \quad t_{n-1}, \quad t_1^{-1}, \quad \ldots, \quad t_{n-1}^{-1}.$$

Any backtrackless tailless cycle on X is uniquely (up to a starting point on the tree between the last and first e_k) determined by the ordered sequence of e_k through which it passes. In particular, if e_i and e_j are two consecutive e_k in this sequence then the part of the cycle between e_i and e_j is the unique backtrackless path on T joining the last vertex of e_i to the first vertex of e_j. For such e_i and e_j, we know that e_j is not the inverse of e_i as the cycle is backtrackless. Nor is the last edge the inverse of the first. Conversely, if we are given any ordered sequence of edges from the e_k in which no two consecutive edges are inverses of each other and in which the last edge is not inverse to the first edge, there is a unique (up to a starting point on the tree between the last and first e_k) backtrackless tailless cycle on X whose sequence of e_k is the given sequence.

The free group of rank r generated by the e_k puts a group structure on backtrackless tailless cycles which is equivalent to the fundamental group of X.

When dealing with the fundamental group of X, any closed path starting at a fixed vertex v_0 on X is completely determined up to homotopy by the ordered sequence of e_k through which it passes. If we do away with backtracking, such a path will be composed of a tail on the tree and then a backtrackless tailless cycle corresponding to the same sequence of e_k, followed by the original tail in the reverse direction and ending at v_0 again. Thus the free group of rank r generated by the e_k is identified with the fundamental group of X. We will therefore refer to the free group generated by the e_k as the **fundamental group** of X.

There are two elementary reduction operations for paths written down in terms of directed edges, just as there are elementary reduction operations for words in the fundamental group of X (see Section 2.4). This means that if $a_1, \ldots, a_s$ and e are taken from the e_k and their inverses, the **two elementary reduction operations** are

(1) $a_1 \cdots a_{i-1} e e^{-1} a_{i+2} \cdots a_s \cong a_1 \cdots a_{i-1} a_{i+2} \cdots a_s$
(2) $a_1 \cdots a_s \cong a_2 \cdots a_s a_1$.

Using the first of these operations, each equivalence class of words corresponds to a group element and a word of minimum length in an equivalence class is a **reduced** word in group theory language. Since the second operation is equivalent to conjugation by a_1, an equivalence class using both elementary reductions corresponds to a conjugacy class in the fundamental group. A word of minimum length using both elementary operations corresponds to a word of minimum length in a conjugacy class in the fundamental group. If $a_1, \ldots, a_s$ are taken from $e_1, \ldots, e_{2r}$ then a word $C = a_1 \cdots a_s$ is of minimum length in its conjugacy class iff $a_{i+1} \neq a_i^{-1}$ for $1 \leq i \leq s-1$ and $a_1 \neq a_s^{-1}$. This is equivalent to saying that the word C corresponds to a **backtrackless tailless** cycle. Equivalent cycles correspond to conjugate elements of the fundamental group. A conjugacy class $[C]$ is **primitive** if a word of minimal length in $[C]$ is not a power of another word. We will say that a word of minimal length in its conjugacy class is **reduced in its conjugacy class**. From now on, we will assume that a representative element of $[C]$ is chosen which is reduced in $[C]$.

Definition 12.1 The $2r \times 2r$ **path matrix** Z has ef entry given by the complex variable z_{ef} if $e \neq f^{-1}$ and by 0 if $e = f^{-1}$; here e, f are edges left out of the spanning tree T.

Note that the path matrix Z has only one zero entry in each row, unlike the edge matrix W from Definition 11.2, which is rather sparse unless the graph is a bouquet of loops. Next we imitate the definition of the edge zeta function.

Definition 12.2 Define the **path norm** for a path $C = a_1 \cdots a_s$ reduced in its conjugacy class $[C]$ and for which $a_i \in \left\{e_1^{\pm 1}, \ldots, e_r^{\pm 1}\right\}$ as

$$N_{\mathrm{P}}(C) = z_{a_1 a_2} \cdots z_{a_{s-1} a_s} z_{a_s a_1}.$$

Then the **path zeta function** is defined for small $|z_{ij}|$ to be

$$\zeta_{\mathrm{P}}(Z, X) = \prod_{[C]} (1 - N_{\mathrm{P}}(C))^{-1},$$

where the product is over primitive reduced conjugacy classes $[C]$ other than the identity class.

We have similar results to those for the edge zeta.

Theorem 12.3 (Determinant formula for path zeta)

$$\zeta_{\mathrm{P}}(Z, X)^{-1} = \det(I - Z).$$

Proof **(Exercise)** Imitate the proof of Theorem 11.4 for the edge zeta. □

Next we want to find a way to obtain the edge zeta from the path zeta. To do this requires a procedure called **specializing the path matrix to the edge matrix.** Use the notation above for the edges e_i left out of the spanning tree T and the edges t_j of T. A closed backtrackless tailless path C is first written as a product of generators of the fundamental group and then as a product of actual edges e_i and t_k. Do this by inserting $t_{k_1} \cdots t_{k_s}$, which is the unique non-backtracking path on T joining the terminal vertex of e_i and the starting vertex of e_j if e_i and e_j are successive deleted or cut edges in C. Now **specialize the path matrix Z to $Z(W)$ with entries**

$$z_{ij} = w_{e_i t_{k_1}} w_{t_{k_1} t_{k_2}} \cdots w_{t_{k_{s-1}} t_{k_s}} w_{t_{k_s} e_j}. \tag{12.1}$$

Then the path zeta function at $Z(W)$ specializes to the edge zeta function.

Theorem 12.4 *Using the specialization procedure defined above, we have*

$$\zeta_{\mathrm{P}}(Z(W), X) = \zeta_E(W, X).$$

Proof The result should be clear since the two defining infinite products coincide. □

Horton [57] has a Mathematica program to do the specialization in formula (12.1).

Note that $\zeta_P(Z, X) = \zeta_E(Z, X^\#)$, where $X^\#$ is the graph obtained from X by fusing all the edges of the spanning tree T to a point. Thus $X^\#$ consists of a bouquet of r loops as in Figure 2.3.

Example 12.5: The dumbbell again Recall that the edge zeta of the dumbbell graph of Figure 11.1 was evaluated by a 6×6 determinant. The path zeta requires a 4×4 determinant. Take the spanning tree to be the vertical edge; this is really the only choice here. One finds using the determinant formula for the path zeta and the specialization of the path zeta to the edge zeta:

$$\zeta_E(W, X)^{-1} = \det \begin{pmatrix} w_{11} - 1 & w_{12}w_{23} & 0 & w_{12}w_{26} \\ w_{35}w_{51} & w_{33} - 1 & w_{35}w_{54} & 0 \\ 0 & w_{42}w_{23} & w_{44} - 1 & w_{42}w_{26} \\ w_{65}w_{51} & 0 & w_{65}w_{54} & w_{66} - 1 \end{pmatrix}. \tag{12.2}$$

If we shrink the vertical edge to a point (which we call **fusion** or contraction), the edge zeta of the new graph is obtained by replacing any products $w_{x2}w_{2y}$ (for $x, y = 1, 3, 4, 6$) which appear in formula (12.2) by w_{xy} and any products $w_{x5}w_{5y}$ (for $x, y = 1, 3, 4, 6$) by w_{xy}. This gives the zeta function of the new graph obtained from the dumbbell by fusing the vertical edge.

Example 12.6 The path zeta function of the tetrahedron specializes to the edge zeta function of the tetrahedron.

Refer to Figure 12.1 and label the inverse edges with the corresponding capital letters. List the edges that index the entries of the matrix Z as a, b, c, A, B, C. You will then find that the matrix $Z(W)$ for the tetrahedron is

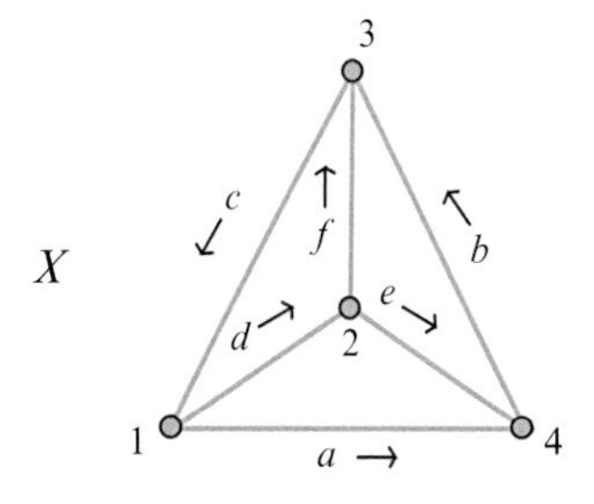

Figure 12.1 Labeling the edges of the tetrahedron.

$$\begin{pmatrix} w_{aE}w_{ED}w_{Da} & w_{ab} & w_{aE}w_{Ef}w_{fc} & 0 & w_{aE}w_{Ef}w_{fB} & w_{aE}w_{ED}w_{DC} \\ w_{bF}w_{FD}w_{Fa} & w_{bF}w_{Fe}w_{eb} & w_{bc} & w_{bF}w_{Fe}w_{eA} & 0 & w_{bF}w_{FD}w_{DC} \\ w_{ca} & w_{cd}w_{de}w_{eb} & w_{cd}w_{df}w_{fc} & w_{cd}w_{de}w_{eA} & w_{cd}w_{df}w_{fB} & 0 \\ 0 & w_{Ad}w_{de}w_{eb} & w_{Ad}w_{df}w_{fc} & w_{Ad}w_{de}w_{eA} & w_{Ad}w_{df}w_{fB} & w_{AC} \\ w_{BE}w_{ED}w_{Da} & 0 & w_{BE}w_{Ef}w_{fc} & w_{BA} & w_{BE}w_{Ef}w_{fB} & w_{BE}w_{ED}w_{DC} \\ w_{CF}w_{FD}w_{Da} & w_{CF}w_{Fe}w_{eb} & 0 & w_{CF}w_{Fe}w_{eA} & w_{CB} & w_{CF}w_{FD}w_{DC} \end{pmatrix}.$$

Exercise 12.1 As a check on the preceding example, specialize all the variables in the $Z(W)$ matrix to $u \in \mathbb{C}$ and call the new matrix $Z(u)$. Check that $\det(I - Z(u))$ is the reciprocal of the Ihara zeta function $\zeta_X(u)$.

Exercise 12.2 Compute the path zeta function for your favorite graph.

Part IV

Finite unramified Galois coverings of connected graphs

Once again, we assume the usual hypotheses for all graphs. These hypotheses were stated in Section 2.1. The unweighted graph X has vertex set V and (undirected) edge set E. It is possibly irregular and possibly has loops and multiple edges. We view a graph covering as an analog of an extension of algebraic number fields or function fields. It is also an analog of a covering of Riemann surfaces. Coverings of weighted graphs have been considered by Chung and Yau [29] as well as Osborne and Severini [96]. The latter paper applies graph coverings to quantum computing.

All the coverings considered here will be unramified unless stated otherwise. In Section 13.3 we will consider some ramified graph coverings.

13
Finite unramified coverings and Galois groups

In this chapter we begin the study of Galois theory for finite unramified covering graphs. It leads to a generalization of Cayley and Schreier graphs and provides factorizations of zeta functions of normal coverings into products of Artin L-functions associated with representations of the Galois group of the covering. Coverings can also be used in constructions of Ramanujan graphs and in constructions of pairs of graphs that are isospectral but not isomorphic. Most of this chapter is taken from Stark and Terras [120]. Other references are Sunada [128] and Hashimoto [51]. Another theory of graph covering which is essentially equivalent can be found in Gross and Tucker [47]. Our coverings, however, differ in that we require all our graphs to be connected and in that our aim is to find analogs of the basic properties of finite-degree extensions of algebraic number fields and their zeta functions. It is also possible to consider infinite coverings such as the universal covering tree T of a finite graph X. We will not do so here except in passing. This is mostly a book about finite graphs, after all.

13.1 Definitions

If our graphs had no multiple edges or loops, our definition of covering would be Definition 2.9. If we want to prove the fundamental theorem of Galois theory for graphs with loops and multiple edges, Definition 2.9 will not be sufficient. We need to produce a more complicated definition of graph covering involving neighborhoods in directed graphs. This definition is necessary for the proof of the unique lifting property; see Proposition 13.3. See Figure 13.7 for an example illustrating the need for our definitions from the point of view of developing the fundamental theorem of Galois theory. See also Massey [83], p. 201, for the same definitions. First we need to think about directed coverings.

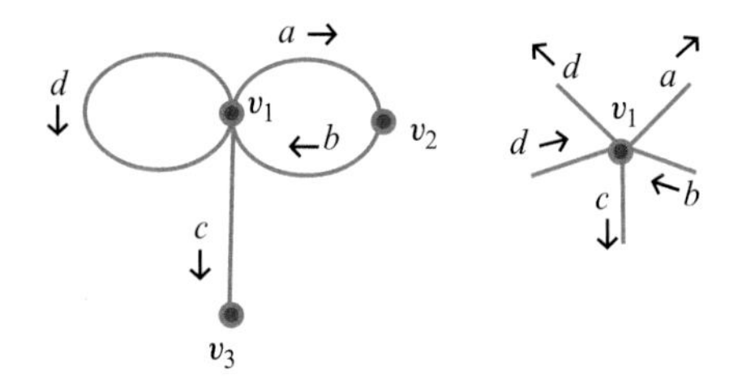

Figure 13.1 *Left*, a directed graph; *right*, a neighborhood of a vertex v_1.

The one-third in the following definition of (directed) neighborhood could be replaced by any $\varepsilon > 0$.

Definition 13.1 A **neighborhood** N of a vertex v in a directed graph X is obtained by taking one-third of each edge at v. The labels and directions are to be included. See Figure 13.1.

Definition 13.2 An undirected finite graph Y is a **covering** of an undirected graph X if, after arbitrarily directing the edges of X, there is an assignment of directions to the edges of Y and an onto **covering map** $\pi : Y \longrightarrow X$ sending neighborhoods of Y one-to-one onto neighborhoods of X which preserve directions.

Note that a covering map π not only takes vertices of Y to vertices of X but also edges of Y to edges of X. The fact that Y is a covering of X is independent of the choice of directions on X. In coverings Y over X (written Y/X) involving loops and multiedges, it is useful to **label** the edges of X and then give the edges of the cover analogous labels in order to see that we really have a covering map. We have attempted to do this for all the examples that follow. See Figure 13.7.

See Figure 13.2 for an example of an invalid assignment of directions in Y over X. Note also that if you create a connected covering of a graph with loops (see Figure 13.7) you may get a graph with multiple edges. Thus, once you allow loops you cannot discuss the general covering without allowing multiple edges. The example in Figure 13.2 is an illegal covering map since a neighborhood of vertex v of X has one edge going in and one going out (once you take one-third of each edge), while that is not true for the neighborhoods of v' and v'' in Y.

If Y covers X, with covering map π, we say that a subgraph $\widetilde{G}$ of Y is a **lift** of a subgraph G of X if $\pi(\widetilde{G}) = G$ and $\widetilde{G}$ and G have the same number of edges.

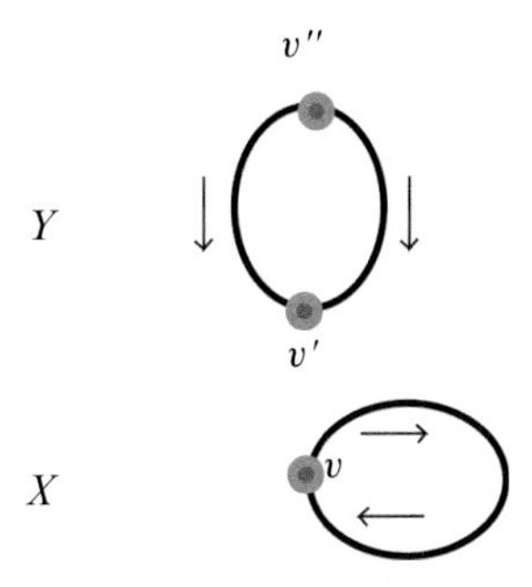

Figure 13.2 This is an example of an invalid assignment of directions in Y/X (i.e., it does not satisfy the rules given in Definition 13.2), since a neighborhood of vertex v of X has one edge going in and one going out (once you take one-third of each edge), while that is not true for the neighborhoods of v' and v''.

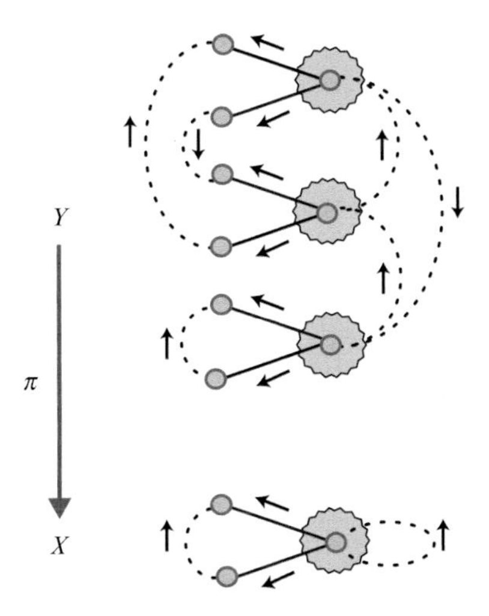

Figure 13.3 A three-sheeted covering. The shaded "fuzzy" area in X is a neighborhood of a selected vertex. The three fuzzy areas in Y are the inverse images under π of the fuzzy area in X.

We construct a covering Y of a connected graph X, as follows. First we find a spanning tree T in X. For a d-sheeted covering, make d copies of T. This gives the nd vertices of our graph Y. That is, Y can be viewed as the set of points (x, i), $x \in X$, $i = 1, \ldots, d$. Then lift to Y the edges of X left out of T to get edges of Y i.e., create edges $\tilde{e}$ of Y with $\pi(\tilde{e}) = e$, starting on each spanning tree copy making up Y. See Figure 13.3.

We regard the copies of the spanning tree as the **sheets** of the covering Y of X. Thus the cube (see Figure 13.4) is a 2-sheeted covering of the tetrahedron.

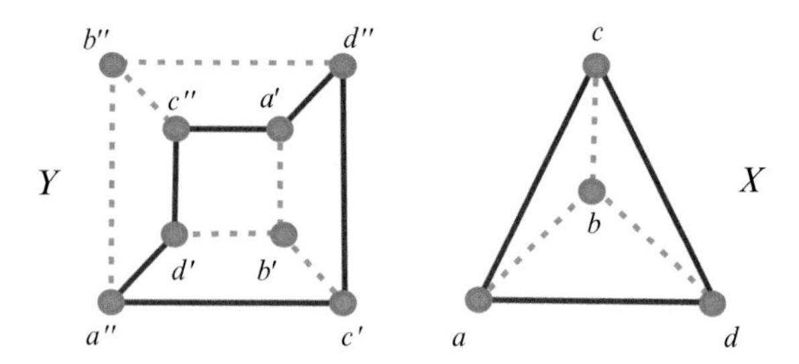

Figure 13.4 The cube is a normal quadratic covering of the tetrahedron. The two sheets of Y are copies of the spanning tree in X (dotted lines).

We refer to such a 2-sheeted covering as a **quadratic** covering in keeping with the terminology from number fields. Similarly we call a 3-sheeted covering **cubic**. A 4-sheeted covering is **quartic**, and so on.

Conversely, as in the proof of Proposition 13.3 below, we see that the spanning tree in X has a unique lift intersecting each vertex in $\pi^{-1}(v)$, for a fixed vertex v of X. This gives the sheets of the cover.

We need to recall a result from topology about the uniqueness of liftings of paths C in X to a unique path starting on sheet 1, say, in a covering space Y. See Massey [83], pp. 151 and 201, for example. As noted above, we have constructed our definitions of coverings involving directed neighborhoods in such a way that this result works even in the presence of loops and multiple edges.

Proposition 13.3 (Uniqueness of lifts of paths in covers) *Suppose that Y is a covering of X. Let C be a path in X. Then C has a unique lift to $\widetilde{C}$, a path in Y, once the initial vertex of $\widetilde{C}$ is fixed.*

Proof Let π be the covering map from Y onto X. According to our definitions we may assume that every edge of Y and X is directed and that π preserves directions. Suppose that e is a directed edge starting at vertex a in X. Then e has a unique lift to $\tilde{e}$ in Y such that $\pi(\tilde{e}) = e$ once you know which vertex in $\pi^{-1}(a)$ is the starting vertex for $\tilde{e}$. Figure 13.7 shows many examples of such lifts.

Now, to obtain the unique lift of a path $e_1 e_2 \cdots e_s$ in X, you just lift each directed edge e_j in order, as j goes from 1 to s, completing the proof. □

Exercise 13.1 Consider any of the coverings in Figure 13.7. Show that Proposition 13.3 is false if we delete the directions on edges.

Next we want to define what we mean by a Galois or normal covering. Of course this will be our favorite kind of cover, since our aim is to develop Galois theory for coverings.

Definition 13.4 If Y/X is a d-sheeted covering with projection map $\pi : Y \longrightarrow X$, we say that it is a **normal or Galois covering** when there are d graph automorphisms $\sigma : Y \longrightarrow Y$ such that $\pi \circ \sigma = \pi$. The **Galois group** $G(Y/X)$ is the set of the maps σ. By a **graph automorphism** we mean a one-to-one onto map of vertices and directed edges of Y that preserves directions.

Later we will see that if we want to make Y a normal cover of X, with Galois group G, we can make use of an appropriate permutation representation π of G to tell us how to lift edges. Let us consider a few basic examples. We will explain how to construct such examples once we have the basic facts about the Galois theory of normal graph coverings.

Example 13.5: The cube as a normal quadratic or 2-sheeted covering of the tetrahedron See Figure 13.4, where the edges in a spanning tree for X are shown as dotted lines. The edges of the corresponding two sheets of Y are also shown as dotted lines.

Exercise 13.2 Create another 2-cover Y' of K_4 using the same spanning tree as in Figure 13.4, except this time when you lift the three non-tree edges of K_4, arrange things so that the lift of only one non-tree edge goes from sheet 1 to sheet 2 while the other two lifts of non-tree edges do not change sheet. Is Y' normal over K_4?

Example 13.6: A non-normal cubic covering of K_4 See Figure 13.5.

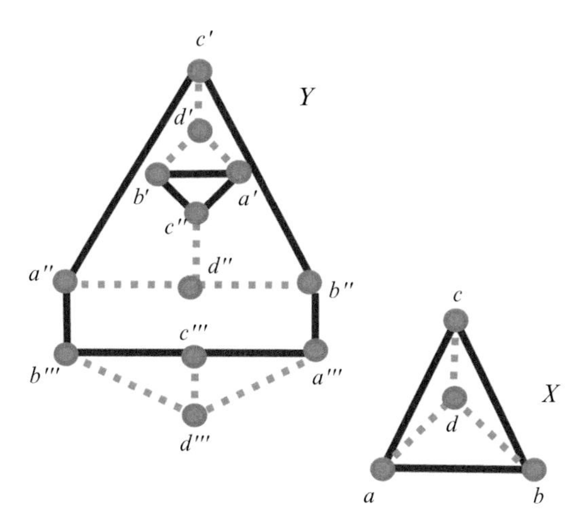

Figure 13.5 A non-normal cubic (3-sheeted) covering of the tetrahedron. The spanning tree in $X = K_4$ is shown with dotted lines, as are the sheets of the covering Y.

Exercise 13.3 Explain why the 3-sheeted covering in Figure 13.5 is not a normal covering of the tetrahedron. *Hint:* a' is adjacent to b' but a'' is not adjacent to b''.

Proposition 13.7 *Suppose that Y/X is a normal covering. The Galois group* $G = G(Y/X)$ *acts transitively on the sheets of the covering.*

Proof By Proposition 13.3, each spanning tree has a unique lift starting at any point in $\pi^{-1}(v_0)$, where v_0 is a fixed point in X. These d lifts are the sheets of the covering.

An automorphism $\sigma \in G$ that fixes a vertex $\widetilde{v_0} \in \pi^{-1}(v_0)$ is the identity. To see this, suppose that $\tilde{v}$ is any vertex in Y. Then a path $\widetilde{P}$ from $\widetilde{v_0}$ to $\tilde{v}$ in Y projects under π to a path from v_0 to $v = \pi(\tilde{v})$ in X. So $\sigma(\widetilde{P})$ is also a lift of P starting at $\widetilde{v_0}$. Thus, by Proposition 13.3 we see that $\sigma(\widetilde{P}) = \widetilde{P}$. It follows, since $\tilde{v}$ was arbitrary, that σ must be the identity.

So, each distinct automorphism $\sigma \in G$ takes $\widetilde{v_0}$ to a different point and there are only d different points in Y above v_0. It follows that the action of G is transitive. Otherwise two different automorphisms would take v_0 to the same point and we have just shown that this is impossible. □

Notation 13.8 (Our notation for vertices and sheets of a normal cover) Suppose that Y/X is normal with Galois group G. We will choose one of the sheets of Y and call it sheet 1. The image of sheet 1 under an element g in G will be called **sheet** g. Any vertex $\tilde{x}$ on Y can then be uniquely denoted $\tilde{x} = (x, g)$, where $x = \pi(\tilde{x})$ and g is the sheet containing $\tilde{x}$.

Definition 13.9 (Action of the Galois group) The Galois group $G(Y/X)$ moves sheets of Y via $g \circ (sheet\ h) = sheet(gh)$:

$$g \circ (x, h) = (x, gh), \qquad \text{for } x \in X, \qquad g, h \in G.$$

It follows that g moves a path in Y as follows:

$$g \circ (\text{path from } (a, h) \text{ to } (b, j)) = \text{path from } (a, gh) \text{ to } (b, gj). \qquad (13.1)$$

Even non-normal coverings Y over X have the nice property that the base graph, $\zeta(u, X)^{-1}$, divides the covering $\zeta(u, Y)^{-1}$, as the following proposition shows. The analogous fact is only conjectured for Dedekind zetas of extensions of number fields.

Proposition 13.10 (Divisibility properties of zeta functions of covers) *Suppose that Y is a d-sheeted (possibly non-normal) covering of X. Then* $\zeta(u, X)^{-1}$ *divides* $\zeta(u, Y)^{-1}$.

Proof Start with the Ihara formula $\zeta(u,Y)^{-1} = (1-u^2)^{r_Y-1}\det(I_Y - A_Y u + Q_Y u^2)$. Note that $r_Y - 1 = |E_Y| - |V_Y| = d\,(|E_X| - |V_X|)$. Thus $(1-u^2)^{r_X-1}$ divides $(1-u^2)^{r_Y-1}$.

Now order the vertices of Y in blocks corresponding to the sheets of the cover, so that A_Y consists of blocks $\widetilde{A}_{ij}$, with $1 \le i, j \le d$, such that $\sum_j \widetilde{A}_{ij} = A_X$.

The same ordering puts Q_Y into block diagonal form, with d copies of Q_X down the diagonal. Similarly I_Y has block diagonal form consisting of d copies of I_X down the diagonal.

Consider $I_Y - A_Y u + Q_Y u^2$. Without changing the determinant, we can add to the first block column the $d-1$ block columns to its right. The new first column is

$$\begin{pmatrix} I_X - A_X u + Q_X u^2 \\ \vdots \\ I_X - A_X u + Q_X u^2 \end{pmatrix}.$$

Now subtract the first block row from all the rest of the block rows. Then the first block column becomes

$$\begin{pmatrix} I_X - A_X u + Q_X u^2 \\ 0 \\ \vdots \\ 0 \end{pmatrix}.$$

The proposition follows. □

13.2 Examples of coverings

This section should provide enough examples to clarify the definitions in Section 13.1.

Example 13.11: An n-cycle is a normal n-fold covering of a loop with cyclic Galois group See Figure 13.6 for this example.

The Ihara zeta function of the loop X in Figure 13.6 is $\zeta_X(u) = (1-u)^{-2}$, and the zeta function of the n-cycle is $\zeta_Y(u) = (1-u^n)^{-2}$. Thus

$$\zeta_Y(u) = (1-u^n)^{-2} = \prod_{j=0}^{n-1} (1 - w^j u)^{-2}, \qquad \text{where } w = e^{2\pi i/n}.$$

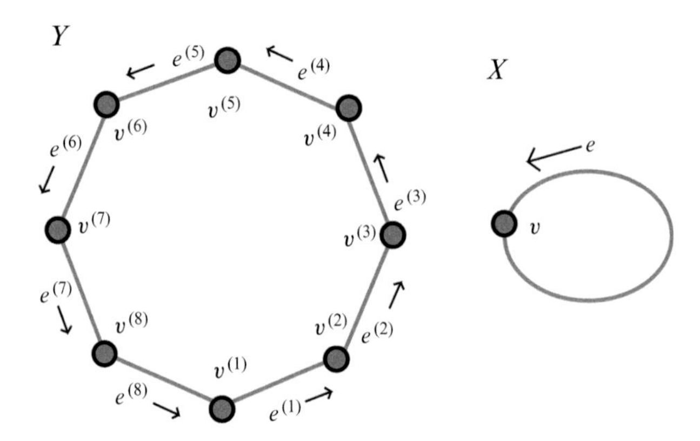

Figure 13.6 An n-cycle is a normal n-fold covering of a loop with cyclic Galois group.

This factorization of $\zeta_Y(u)$ will later be seen as a special case of the factorization of the Ihara zeta function of Y into a product of Artin L-functions associated with representations of the Galois group of Y/X. See Corollary 18.11.

Question Suppose a graph Y has a large symmetry group S and that G is a subgroup of S. Is there a graph X such that Y is a normal cover of X with group G?

Answer Not always. For example, the cube has symmetry group S_4 – a group of order 24. But if G is a subgroup of order d such that d does not divide 4, G cannot be the Galois group $G(Y/X)$. Why? If Y/X were a Galois cover with d sheets, it would follow that d divides the number of vertices (and edges) of Y. But the cube has eight vertices and 12 edges. Therefore d must divide $r_Y - 1 = |E| - |V| = 12 - 8 = 4$.

The next figure shows why our definitions were so messy.

Example 13.12: A Klein 4-group ($\mathbb{Z}_2 \times \mathbb{Z}_2$)-cover of the dumbbell This illustrates that we need Definition 13.2 and our definition, to be given in Chapter 14, of an intermediate cover in order to obtain the fundamental theorem of Galois theory. See Figure 13.7 for this example. It is not hard to check that Z is a Galois cover of the dumbbell X with Galois group $\mathbb{Z}_2 \times \mathbb{Z}_2$. In the next chapter we will see that according to our definitions of intermediate cover, we have three intermediate 2-covers Y, Y'', Y'''. The last two are isomorphic if you ignore directions. That would mean that ignoring directions invalidates the fundamental theorem of Galois theory.

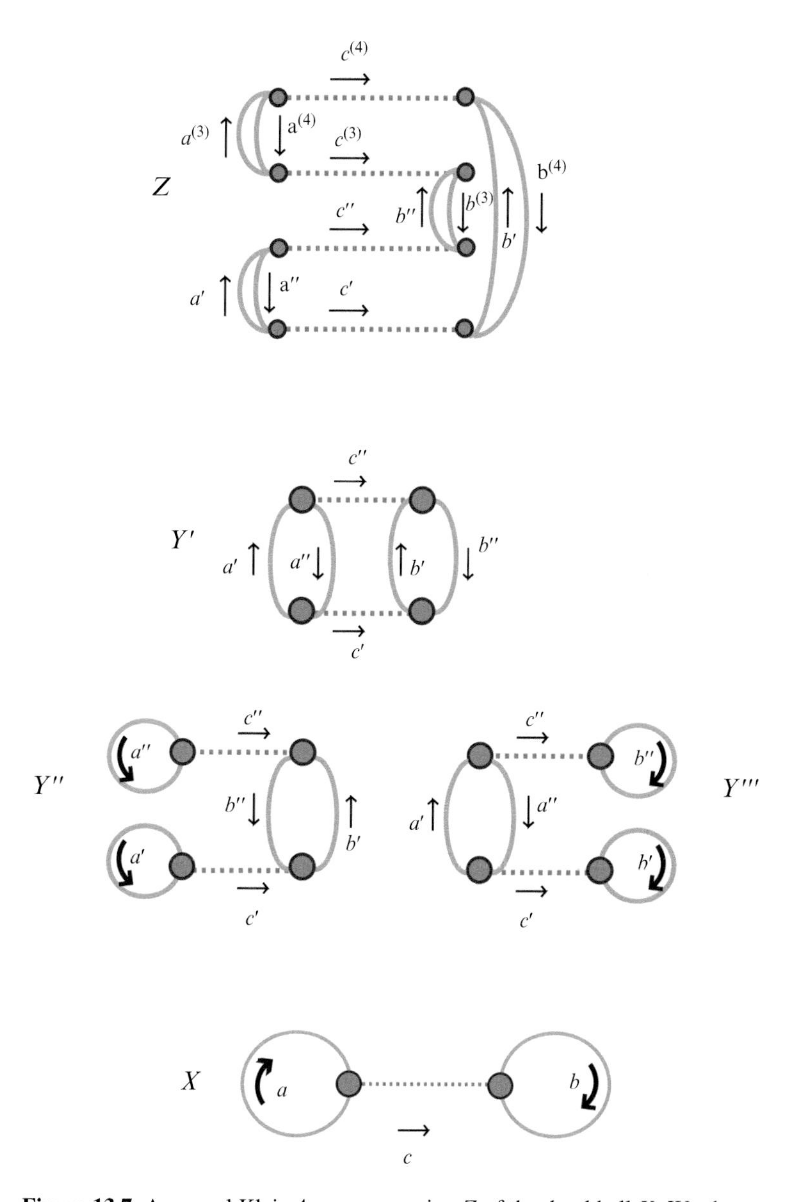

Figure 13.7 A normal Klein 4-group covering Z of the dumbbell X. We show three intermediate 2-covers of X, i.e., Y, Y', and Y'''. Note that the last two are isomorphic as undirected graphs. The spanning tree of X and the sheets of the covers are indicated by dotted lines.

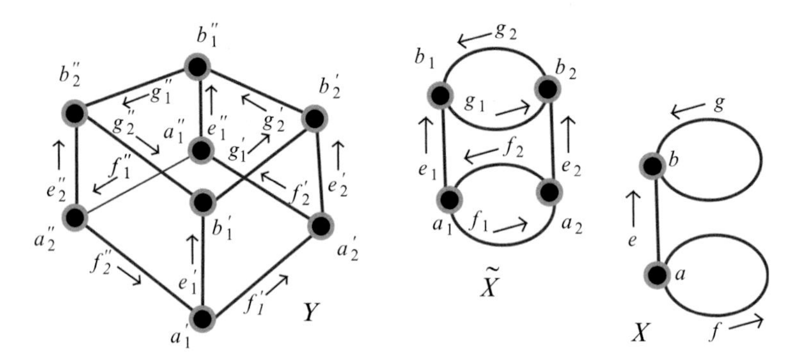

Figure 13.8 An order-4 **cyclic cover** Y/X, where Y is the cube. Included is the intermediate quadratic cover $\widetilde{X}$. The notation should make clear the covering projections $\pi : Y \longrightarrow X,\ \pi_2 : Y \longrightarrow \widetilde{X},\ \pi_1 : \widetilde{X} \longrightarrow X$.

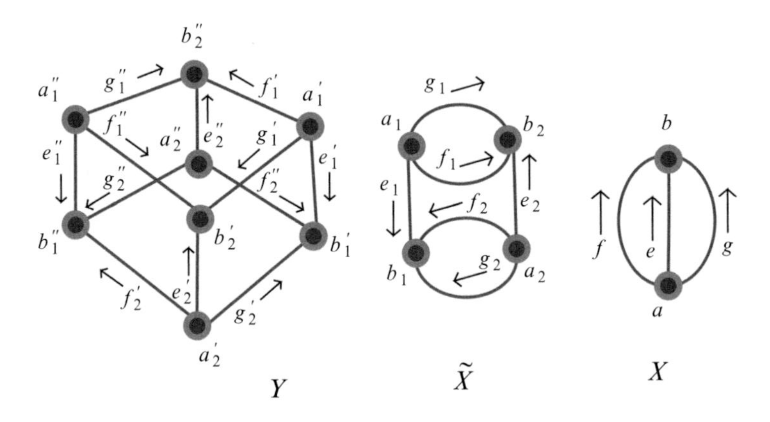

Figure 13.9 A Klein 4-group cover Y/X, where Y is the cube. Included is $\widetilde{X}$, one of the three intermediate quadratic covers.

Example 13.13: Two graphs with the cube as a normal cover See Figures 13.8 and 13.9 for these examples. Let Y be the cube. Then $|V| = 8$, $|E| = 12$, and $|G|$ divides g.c.d.$(8, 12) = 4$. Figure 13.8 shows a normal covering Y/X where Y is the cube, such that $G = G(Y/X)$ is a cyclic group of order 4. Figure 13.9 shows another such covering Y/X, where $G = G(Y/X)$ is the Klein 4-group. In each case we include an intermediate quadratic cover in the figures. The concept of an intermediate cover will be discussed in Chapter 14.

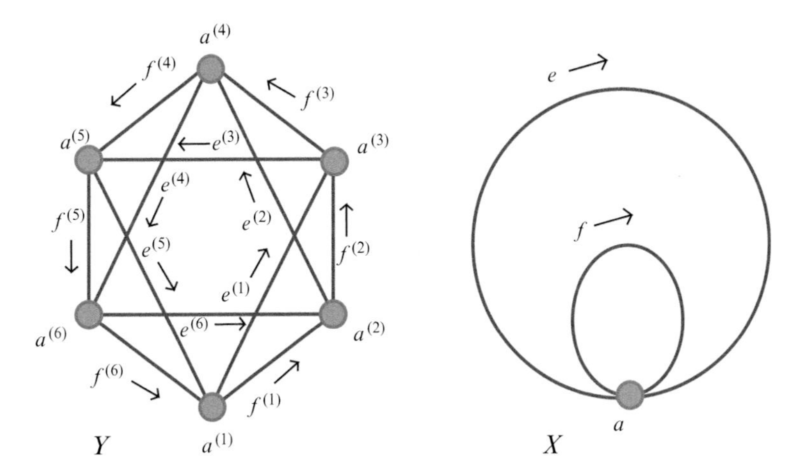

Figure 13.10 A cyclic 6-fold cover Y/X, where Y is the octahedron.

Example 13.14: The octahedron as a cyclic 6-fold cover of two loops The octahedron has $|V| = 6$, $|E| = 12$ which implies that a Galois group of order 6 is possible for the octahedron as a covering of a bouquet of two loops. See Figure 13.10 for an example.

13.3 Some ramification experiments

The word "ramified" comes from the theory of extensions of algebraic number fields or function fields over finite fields. In particular, there is a **conjecture of Dedekind** saying that if K is an extension of the number field F then the Dedekind zeta function $\zeta_K(s)$ divides $\zeta_K(s)$ even when there is ramification. One can also view the theory of graph coverings as analogous to coverings of Riemann surfaces or topological manifolds. A ramified surface would be a branched surface such as the Riemann surface of $\sqrt{z}$ or $\Gamma\backslash H$ with a discrete group Γ acting with fixed points, such as the modular group $\mathrm{SL}(2, \mathbb{Z})$. I have tried some experiments on ramifying the vertices and edges of coverings. See the examples below. References are Beth Malmskog and Michelle Manes [81] and also Baker and Norine [8].

Ramified example 1 The zeta function of a graph L_n consisting of one vertex and n loops can be found from the Ihara determinant formula in Theorem 2.5. Here the adjacency matrix $A = 2n$ is 1×1. The matrix $Q = 2n - 1$ is 1×1 also. The rank of the fundamental group is $r = n$. The Ihara formula says that

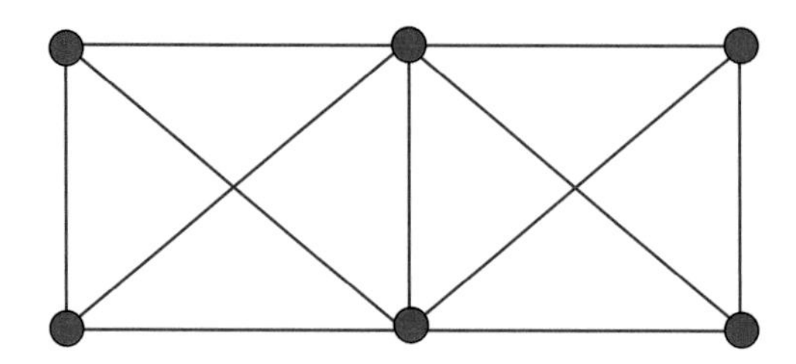

Figure 13.11 Edge ramified cover Z of K_4, the complete graph on four vertices, obtained by taking two copies of K_4 and identifying an edge.

$$\begin{aligned}\zeta(u, L_n)^{-1} &= (1-u^2)^{r_X-1}\det(I - A_X u + Q_X u^2)\\ &= (1-u^2)^{n-1}(1-2nu+(2n-1)u^2)\\ &= (1-u^2)^{n-1}(1-u)(1-(2n-1)u).\end{aligned}$$

If we view L_n as a ramified covering of L_1, we are happy since $\zeta(u, L_1)^{-1}$ divides $\zeta(u, L_n)^{-1}$. However, $\zeta(u, L_2)^{-1}$ does not divide $\zeta(u, L_{2n})^{-1}$. The good thing is that there is only one bad factor, $[1-(2n-1)u]$, and it is linear.

It is easy to turn this example into a bouquet of n triangles T_n covering a triangle T_1. Since each path in L_n is tripled in length, we see that $\zeta(u, T_n) = \zeta(u^3, L_n)$. The good thing is that still only one factor is bad. See Malmskog and Manes [81] for a more general result.

Ramified example 2 Next we obtain a graph Z from two copies of K_4, identifying (fusing) an edge. See Figure 13.11. Here I was looking for an analog of the Riemann surface of $\sqrt{z}$.

We compute the zeta function of Z in Figure 13.11 to be

$$\begin{aligned}\zeta(u, Z)^{-1} &= (1-u^2)^5(u-1)(4u^2+u+1)\\ &\quad \times (2u^2-u+1)(8u^3+2u^2+u-1)(2u^2+u+1)^2.\end{aligned}$$

The only factor of $\zeta(u, K_4)^{-1}$ that does not divide $\zeta(u, Y)^{-1}$ is $1-2u$. This looks again like the result of Beth Malmskog and Michelle Manes [81].

Exercise 13.4 Compute the zeta when you identify or fuse an edge on n copies of K_4. Consider the divisibility properties of the corresponding zetas.

14

Fundamental theorem of Galois theory

Question What does it mean to say that $\widetilde{X}$ is intermediate to Y/X? Our goal is to prove the fundamental theorem of graph Galois theory, e.g., the existence of a one-to-one correspondence between subgroups H of the Galois group G of Y/X and graphs $\widetilde{X}$ intermediate to Y/X. For this we need a definition which is stronger than just saying $Y/\widetilde{X}$ is a covering and $\widetilde{X}/X$ is a covering. To see this, consider Figures 13.7 and 13.9. These examples would contradict the fundamental theorem of Galois theory for graph coverings if we were to make over-simplistic definitions. To avoid this problem, we will make a definition that is perhaps annoyingly complicated.

Exercise 14.1 Draw the other two intermediate graphs for Figure 13.9.

Definition 14.1 Suppose that Y is a covering of X with projection map π. A graph $\widetilde{X}$ is an **intermediate covering** to Y/X if $Y/\widetilde{X}$ is a covering, $\widetilde{X}/X$ is a covering and the projection maps $\pi_1 : \widetilde{X} \longrightarrow X$ and $\pi_2 : Y \longrightarrow \widetilde{X}$ have the property that $\pi = \pi_1 \circ \pi_2$.

See Figure 14.1. Technically, it is the triple $(\widetilde{X}, \pi_1, \pi_2)$ that gives the intermediate covering.

A second definition (which may also make the reader's hair stand on end) tells us when two intermediate graphs are to be considered the same or equal. Again, please remember our goal, which is to prove the fundamental theorems of Galois theory.

Definition 14.2 Let $\widetilde{X}$ and $\widetilde{X}'$ be intermediate to Y/X with projection maps as in Figure 14.1. We assume all graphs have edges which have been assigned directions that are consistent with the projection maps. Suppose that i is a graph isomorphism between $\widetilde{X}$ and $\widetilde{X}'$ (meaning it is one-to-one onto on vertices and directed edges). If the notation is as in Figure 14.1, and we have $\pi_1 = \pi_1' \circ i$,

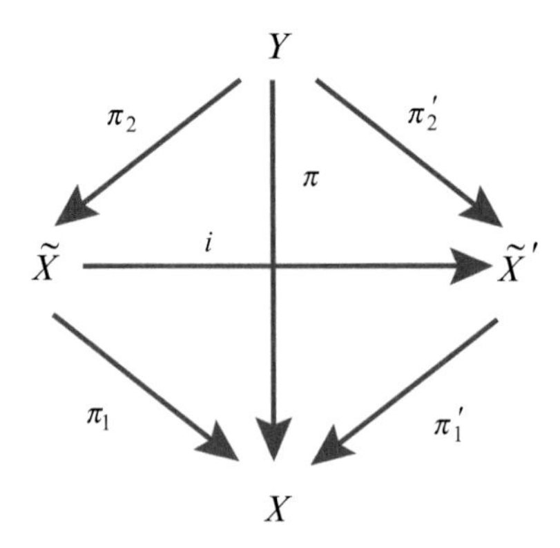

Figure 14.1 A covering isomorphism i of intermediate graphs.

then we say i is a **covering isomorphism** and that $\widetilde{X}$ and $\widetilde{X}'$ are **covering isomorphic**. If, in addition, we have $i \circ \pi_2 = \pi_2'$ then we say that $\widetilde{X}$ and $\widetilde{X}'$ are **the same** or **equal**.

Later in Theorem 14.5 we will see that covering isomorphic graphs intermediate to the Galois cover Y/X with Galois group G correspond to conjugate subgroups of G. This means that covering isomorphic intermediate graphs are analogous to number fields such as $\mathbb{Q}\left(\sqrt[3]{2}\right)$ and $\mathbb{Q}\left(e^{2\pi i/3}\sqrt[3]{2}\right)$.

Now we can prove the fundamental theorem. Note that most of the proofs of the various parts of the theorem are based on the uniqueness of lifts; see Proposition 13.3.

Theorem 14.3 (Fundamental theorem of Galois theory) *Suppose that Y/X is an unramified normal covering with Galois group $G = G(Y/X)$.*

(1) *Given a subgroup H of G, there exists a graph $\widetilde{X}$ intermediate to Y/X such that $H = G(Y/\widetilde{X})$. Write $\widetilde{X} = \widetilde{X}(H)$.*
(2) *Suppose that $\widetilde{X}$ is intermediate to Y/X. Then there is a subgroup $H = H(\widetilde{X})$ of G which equals $G(Y/\widetilde{X})$.*
(3) *Two intermediate graphs $\widetilde{X}$ and $\widetilde{X}'$ are equal (as in Definition 14.2) if and only if $H(\widetilde{X}) = H(\widetilde{X}')$.*
(4) *We have $H(\widetilde{X}(H)) = H$ and $\widetilde{X}(H(\widetilde{X})) = \widetilde{X}$. So we write $\widetilde{X} \leftrightarrow H$ for the correspondence between graphs $\widetilde{X}$ intermediate to Y/X and subgroups H of the Galois group $G = G(Y/X)$.*
(5) *If $\widetilde{X_1} \leftrightarrow H_1$ and $\widetilde{X_2} \leftrightarrow H_2$ then $\widetilde{X_1}$ is intermediate to $Y/\widetilde{X_2}$ iff $H_1 \subset H_2$.*

Proof Part (1) Let H be a subgroup of G. The vertices of Y are of the form (x, g), with $x \in X$ and $g \in G$. Define the vertices of $\widetilde{X}$ to be (x, Hg) for

$x \in X$ and coset $Hg \in H\backslash G$. Create an edge from (a, Hr) to (b, Hs), for $a, b \in X$ and $r, s \in G$, iff there are $h, h' \in H$ such that there is an edge from (a, hr) to $(b, h's)$ in Y.

The edge between (a, Hr) and (b, Hs) in $\widetilde{X}$ is given the label and direction of the projected edge between a and b in X.

Exercise 14.2 Show that $\widetilde{X}$ is well defined, intermediate to Y/X, and connected.

Part (2) Let $\widetilde{X}$ be intermediate to Y/X, with projections $\pi : Y \longrightarrow X, \pi_2 : Y \longrightarrow \widetilde{X}$, and $\pi_1 : \widetilde{X} \longrightarrow X$. Fix a vertex $v_0 \in X$ with $\tilde{\tilde{v}}_0 \in \pi^{-1}(v_0)$ on sheet 1 of Y. That is, $\tilde{\tilde{v}}_0 = (v_0, 1)$ using our labeling of the sheets of Y. Let $\widetilde{v_0} = \pi_2(\tilde{\tilde{v}}_0) \in \widetilde{X}$. Define

$$
\begin{aligned}
H &= \left\{h \in G \,\middle|\, h(\tilde{\tilde{v}}_0) \in \pi_2^{-1}(\widetilde{v_0})\right\} \\
&= \{h \in G \,|\pi_2(v_0, h) = \pi_2(v_0, 1)\}. \qquad (14.1)
\end{aligned}
$$

To see that H is a subgroup of G we need only show that H is closed under multiplication. Let h_1 and h_2 be elements of H. Then, by the definition of H, we have $\pi_2(v_0, h_1) = \pi_2(v_0, h_2) = \pi_2(v_0, 1) = \widetilde{v_0}$.

Let $\tilde{\tilde{p}}_1$ and $\tilde{\tilde{p}}_2$ be paths on Y from $(v_0, 1)$ to the vertices (v_0, h_1) and (v_0, h_2), respectively. Then $\tilde{\tilde{p}}_1$ and $\tilde{\tilde{p}}_2$ project under π_2 to closed paths $\widetilde{p_1}$ and $\widetilde{p_2}$ in $\widetilde{X}$ beginning and ending at $\widetilde{v_0}$. And $\tilde{\tilde{p}}_1$ and $\tilde{\tilde{p}}_2$ project under $\pi = \pi_1 \circ \pi_2$ to closed paths p_1 and p_2 in X beginning and ending at v_0.

By formula (13.1), $h_1 \circ \tilde{\tilde{p}}_2$ starts at (v_0, h_1) and ends at (v_0, h_1h_2). Thus the lift of $\widetilde{p_1}\widetilde{p_2}$ from $\widetilde{X}$ to Y beginning at $(v_0, 1)$, which is the same as the lift of p_1p_2 from X to Y beginning at $(v_0, 1)$, ends at (v_0, h_1h_2). It follows that h_1h_2 is in H and H is a subgroup of G.

Question How does H depend on v_0?

For part (3), see below.

Part (4) Let $\widetilde{X}$ be intermediate to Y/X. We want to prove that $\widetilde{X}(H(\widetilde{X})) = \widetilde{X}$, using the definitions from the proofs of parts (1) and (2) as well as Definition 14.2.

Before attempting to prove this equality, we need to prove a characterization of $\pi_2^{-1}(\tilde{v})$ for any vertex $\tilde{v}$ of $\widetilde{X}$. This says that there is an element $g_v \in G$ such that, if $H(\widetilde{X}) = H$,

$$
\pi_2^{-1}(\tilde{v}) = \{(v, hg_v)|h \in H\}. \qquad (14.2)
$$

Let v_0 be the fixed vertex of X from the definition of H in the proof of part (2). Let $\widetilde{q}$ be a path in $\widetilde{X}$ from $\widetilde{v_0}$ to $\tilde{v}$. There is a lift $\tilde{\tilde{q}}$ of $\tilde{q}$ to Y starting

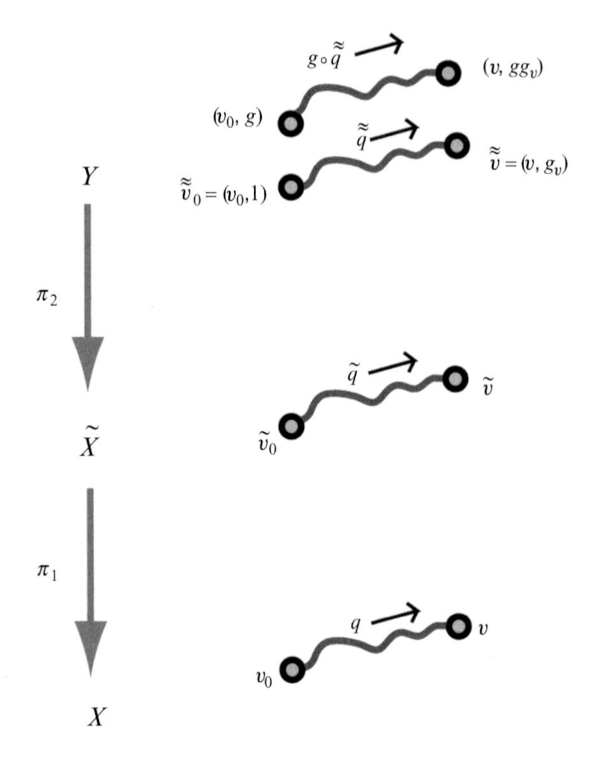

Figure 14.2 Part of the proof of part 4 of Theorem 14.3, showing that $\{(v, hg_0)|h \in H\} = \pi_2^{-1}(\tilde{v})$.

at $(v_0, 1)$ and ending at (v, g_v). Write $\tilde{\tilde{v}} = (v, g_v)$ in Y with $\tilde{v} = \pi_2(\tilde{\tilde{v}}) \in \widetilde{X}$ and $\pi(\tilde{\tilde{v}}) = v$. Projected down to X, we get the path q from v_0 to v.

Look at Figure 14.2. For $g \in G$, by equation (13.1) the path $\widetilde{q}$ in $\widetilde{X}$ lifts to a path $g \circ \tilde{\tilde{q}}$ from (v_0, g) to (v_0, gg_v) in Y. Thus, by the uniqueness of lifts, starting on a given sheet we must have that $\pi_2 \circ g \circ \tilde{\tilde{q}} = \tilde{q}$ if and only if the initial sheet of the lift of q is that of $\widetilde{v_0}$. That is, $\pi_2 \circ g \circ \tilde{\tilde{q}} = \tilde{q}$ iff $g \in H$. This proves formula (14.2).

Now we seek to show that $\widetilde{X}' = \widetilde{X}(H(\widetilde{X})) = \widetilde{X}$ in the sense of Definition 14.2. Recall that $\widetilde{X}' = \widetilde{X}(H(\widetilde{X}))$ has vertices (x, Hg) and projections $\pi_2'(x, g) = (x, Hg)$ and $\pi_1'(x, Hg) = x$. Define $i : \widetilde{X} \to \widetilde{X}'$ by $i(\tilde{v}) = (v, Hg_v)$, using the element $g_v \in G$ from formula (14.2).

Exercise 14.3 (a) Why do $\widetilde{X}' = \widetilde{X}(H(\widetilde{X}))$ and $\widetilde{X}$ have the same number of vertices?

(b) Prove that i is a graph isomorphism (i.e., it is one-to-one onto between vertices and directed edges) and that $i \circ \pi_2 = \pi_2'$ and $\pi_1' \circ i = \pi_1$.

To complete the proof of part (4), we must show that $H(\widetilde{X}(H)) = H$. By the definitions made in the proofs of parts (1) and (2), we have

$$H(\widetilde{X}(H)) = \{g \in G | \pi_2(v_0, g) = \pi_2(v_0, 1)\} = \{g \in G | (v_0, Hg) = (v_0, H)\}$$
$$= H.$$

Part (5) Suppose that $\pi_2 : Y \to \widetilde{X_1}$ and $\pi_1 : \widetilde{X_1} \to \widetilde{X_2}$, with $\pi_3 = \pi_1 \circ \pi_2 : Y \to \widetilde{X_2}$. Then, by the proof of part (2),

$$H(\widetilde{X_1}) = H_1 = \{h \in G | \pi_2(v_0, h) = \pi_2(v_0, 1)\},$$
$$H(\widetilde{X_2}) = H_2 = \{h \in G | \pi_3(v_0, h) = \pi_3(v_0, 1)\}.$$

Since $\pi_3 = \pi_1 \circ \pi_2$ it follows that $H_1 \subset H_2$.

For the converse, suppose that $H_1 \subset H_2$. Then we have intermediate graphs $\widetilde{X_i}$, with vertices $(x, H_i\sigma)$ for $x \in X$, and cosets $H_i\sigma \in H_i\backslash G$. There is an edge between $(a, H_i\sigma)$ and $(b, H_i\tau)$, for $a, b \in X$ and $\sigma, \tau \in G$, iff there are $h, h' \in H_i$ such that $(a, h\sigma)$ and $(b, h'\tau)$ have an edge in Y. We need to show that $\exists\ \pi_2 : Y \to \widetilde{X_1}$ and $\pi_1 : \widetilde{X_1} \to \widetilde{X_2}$ with $\pi_3 = \pi_1 \circ \pi_2 : Y \to \widetilde{X_2}$. Here $\pi_2(v, g) = (v, H_1 g)$ and $\pi_3(v, g) = (v, H_2 g)$ for $v \in X, g \in G$. Then, since $H_1 \subset H_2$, we see that $\pi_1\ (v, H_1 g) = (v, H_2 g)$ makes sense, as $H_1 a = H_1 b$ iff $ab^{-1} \in H_1$. Since $H_1 \subset H_2$, this implies that $H_2 a = H_2 b$.

Part (3) Suppose that we have two graphs $\widetilde{X}$ and $\widetilde{X}'$ intermediate to Y/X with projections $\pi_2 : Y \longrightarrow \widetilde{X}, \pi_1 : \widetilde{X} \longrightarrow X$ and $\pi_2' : Y \longrightarrow \widetilde{X}', \pi_1' : \widetilde{X}' \longrightarrow X$. Set $H = H(\widetilde{X})$ and $H' = H(\widetilde{X}')$. Suppose that $\widetilde{X} = \widetilde{X}'$. Then there is a graph isomorphism $i : \widetilde{X} \to \widetilde{X}'$ as in Definition 14.2 such that $i \circ \pi_2 = \pi_2'$ and $\pi_1' \circ i = \pi_1$. Then

$$H = \{h \in G | \pi_2(v_0, h) = \pi_2(v_0, 1)\},$$
$$H' = \{h \in G | \pi_2'(v_0, h) = \pi_2'(v_0, 1)\}.$$

Since $i \circ \pi_2 = \pi_2'$ and i is one-to-one, we find that $H = H'$.

To go the other way, suppose that $H(\widetilde{X}) = H(\widetilde{X}')$. Then we need to show that there is a graph isomorphism $i : \widetilde{X} \to \widetilde{X}'$ as in Definition 14.2 such that $i \circ \pi_2 = \pi_2'$ and $\pi_1' \circ i = \pi_1$. Note first that $\widetilde{X}$ and $\widetilde{X}'$ have the same number of vertices. We know from formula (14.2) that

$$\pi_2^{-1}(\tilde{v}) = \{(v, hg_v)|h \in H\},$$

$$\pi_2'^{-1}(\widetilde{v'}) = \left\{(v', hg_{v'})|h \in H\right\}.$$

Define $i(\tilde{v})$ as $\pi_2'(v, hg_v)$.

Exercise 14.4 Show that i is a graph isomorphism such that $i \circ \pi_2 = \pi_2'$ and $\pi_1' \circ i = \pi_1$. □

Next we consider **conjugate subgroups** of the Galois group (subgroups H and gHg^{-1} for $g \in G$) and their corresponding intermediate graphs.

Definition 14.4 Suppose we have the following correspondences between intermediate graphs and subgroups of G:

$$\widetilde{X} \longleftrightarrow H \subset G$$

$$\widetilde{X}' \longleftrightarrow gHg^{-1} \subset G \qquad \text{for some } g \in G.$$

We say that $\widetilde{X}$ and $\widetilde{X}'$ are **conjugate**. This definition turns out to be equivalent to part of Definition 14.2.

Theorem 14.5 *Graphs $\widetilde{X}$ and $\widetilde{X}'$ intermediate to the normal cover Y/X with Galois group G are conjugate in the sense of Definition 14.4 if and only if they are covering isomorphic in the sense of Definition 14.2.*

Proof Suppose that H and $H' = g_0 H g_0^{-1}$ are conjugate subgroups of G, where $g_0 \in G$. We want to show that the corresponding intermediate graphs $\widetilde{X} = \widetilde{X}(H)$ and $\widetilde{X}' = \widetilde{X}(H')$ (using the notation of Theorem 14.3) are covering isomorphic in the sense of Definition 14.2. We have the disjoint coset decompositions

$$G = \bigcup_{j=1}^{d} Hg_j \qquad \text{and} \qquad G = \bigcup_{j=1}^{d} H' g_0 g_j.$$

This means that the graphs $\widetilde{X}$ and $\widetilde{X}'$ have vertices $\{(v, Hg_j)|v \in X, 1 \le j \le d\}$ and $\{(v, H'g_0g_j)|v \in X, 1 \le j \le d\}$, respectively. The isomorphism $i : \widetilde{X} \longrightarrow \widetilde{X}'$ is defined by $i(v, Hg) = (v, H'g_0g)$.

Exercise 14.5 Prove that $i : \widetilde{X} \longrightarrow \widetilde{X}'$ is a covering isomorphism in the sense of Definition 14.2.

For the converse, suppose that $\widetilde{X}$ and $\widetilde{X}'$ are covering isomorphic intermediate graphs. We must show that the corresponding subgroups $H = H(\widetilde{X})$ and $H' = H(\widetilde{X}')$ (in the notation of Theorem 14.3) are conjugate. By

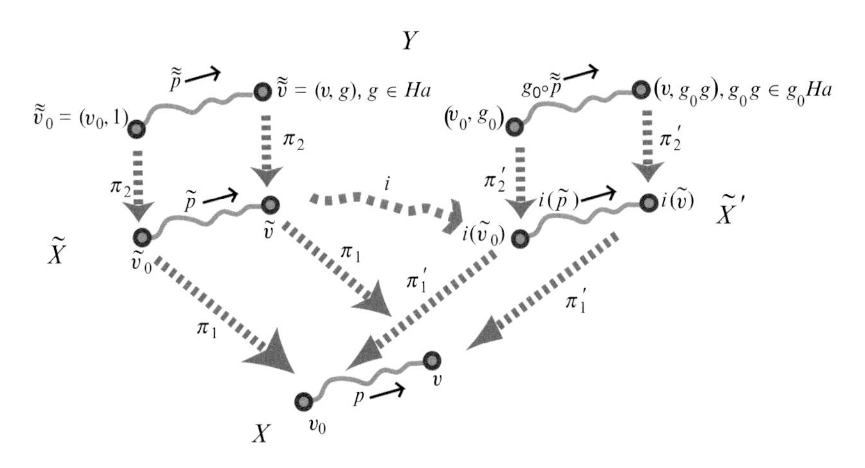

Figure 14.3 Part of the proof of Theorem 14.5.

Definition 14.2, there is an isomorphism $i : \widetilde{X} \longrightarrow \widetilde{X}'$ such that $\pi_1 = \pi_1' \circ i$. Fix vertex $v_0 \in X$ and let $\widetilde{\widetilde{v_0}} = (v_0, 1)$ be on sheet 1 of Y projecting to $\widetilde{v_0} = \pi_2(\widetilde{\widetilde{v_0}})$ in $\widetilde{X}$. For any $\tilde{v} \in \widetilde{X}$, suppose that it projects to $v \in X$ under π_1 and that $\tilde{\tilde{v}} = (v, g_v) \in Y$ projects to $\tilde{v}$ under π_2. See Figure 14.3. The set $\{g \in G | \pi_2(v, g) = \tilde{v}\} = Hg_v$ by formula (14.2). Let $\tilde{\tilde{p}}$ be a path on Y from $\widetilde{\widetilde{v_0}}$ to $\tilde{\tilde{v}}$. It projects via π_2 to a path $\tilde{p}$ in $\widetilde{X}$ from $\widetilde{v_0}$ to $\tilde{v}$ and to a path p in X from v_0 to v.

The path $i(\tilde{p})$ in $\widetilde{X}'$ from $i(\widetilde{v_0})$ to $i(\tilde{v})$ also projects under π_1' to p. As $i(\widetilde{v_0})$ projects under π_1' to v_0, there is a $g_0 \in G$ such that $(v_0, g_0) \in Y$ projects via π_2' to $i(\widetilde{v_0})$. Now $\pi(g_0 \circ \tilde{\tilde{p}}) = \pi(\tilde{\tilde{p}}) = p$. Since $\pi = \pi_1' \circ \pi_2'$, it follows that the path $\pi_2'(g_0 \circ \tilde{\tilde{p}})$ in $\widetilde{X}'$ has initial vertex $i(\widetilde{v_0})$ and projects to p in X. By the uniqueness of lifts, then $i(\tilde{p}) = \pi_2'(g_0 \circ \tilde{\tilde{p}})$. However, $g_0 \circ \tilde{\tilde{p}}$ ends at $(v, g_0 g)$. Therefore π_2' takes $(v, g_0 g)$ to $i(\tilde{v})$. In particular, the set of all such $g_0 g$ is $g_0 H g_v = \left(g_0 H g_0^{-1}\right) g_0 g_v$. Therefore $H' = g_0 H g_0^{-1}$. □

Remark 14.6 The previous proof showed that the effect of the isomorphism i can be achieved by the element $g_0 \in G$. In fact, g_0 may be replaced by any element of the right coset $(g_0 H g_0^{-1}) g_0 = g_0 H$, which is a left coset of H. This gives a **one-to-one correspondence between left cosets $g_0 H$ of H and all possible "embeddings" of $\widetilde{X}$ in** Y/X.

Theorem 14.7 *Suppose that Y/X is a normal covering with Galois group G and $\widetilde{X}$ is an intermediate covering corresponding to the subgroup H of G.*

Then $\widetilde{X}$ is itself a normal covering of X if and only if H is a normal subgroup of G, in which case $G(\widetilde{X}/X) \cong H\backslash G$.

Proof Recall the proof of Theorem 14.5. View $\widetilde{X}$ as $\widetilde{X}(H)$ (using the notation of Theorem 14.3), with vertex set

$$\{(v, Hg_j) | v \in X, 1 \leq j \leq d\},$$

where the g_j are right coset representatives for $H\backslash G$.

Suppose that H is a normal subgroup of G. A coset Hg acts on $\widetilde{X}(H)$ by sending (v, Hg_j) to (v, Hgg_j). This action preserves edges and is transitive on the cosets Hg.

Exercise 14.6 Prove this last statement. You need to use the normality of H to see that the action preserves edges.

This gives $d = |G/H|$ automorphisms of $\widetilde{X}(H)$, showing that $\widetilde{X}(H)$ is normal over X with Galois group G/H.

For the converse, suppose that $\widetilde{X}/X$ is normal and i is an automorphism of $\widetilde{X}$ in $G(\widetilde{X}/X)$. Apply Theorem 14.5 with $\widetilde{X}' = \widetilde{X}$, $\pi_1 = \pi_1'$, and $\pi_2 = \pi_2'$. Although i is not the map that makes $\widetilde{X}' = \widetilde{X}$ (that map is the identity map), nevertheless i is an isomorphism between $\widetilde{X}$ and $\widetilde{X}'$ and it is a conjugation map since $\pi_1' \circ i = \pi_1 \circ i = \pi_1$. Thus Theorem 14.5 says that there is a $g_0 \in G$ such that the intermediate graph $\widetilde{X}'$ corresponds to the subgroup $g_0 H g_0^{-1}$. Since $\widetilde{X}' = \widetilde{X}$, we have $g_0 H g_0^{-1} = H$.

Moreover, if we choose $\widetilde{v_0} \in \widetilde{X}$ as in the proof of Theorem 14.5, we have $\pi_2((v_0, g_0)) = \pi_2'((v_0, g_0)) = i(\widetilde{v_0})$. As i runs through the d elements of $G(\widetilde{X}/X)$, the $i(\widetilde{v_0})$ run through the d lifts of v_0 to $\widetilde{X}$. Thus the corresponding d different g_0 run through the d left cosets of H in G and we have $g_0 H g_0^{-1} = H$ for all these, which says that H is normal in G. □

The reader should now go back and reconsider the examples in Figures 13.8 and 13.9. As an **exercise**, write down all the intermediate graphs. Next let us consider a new example, given in Figure 14.4.

Example 14.8: An S_3 cover of $K_4 - e$ with two intermediate covers Figure 14.4 shows a normal covering Y_6 of $X = K_4 - e$ with Galois group $G(Y_6/X) = S_3$, the symmetric group of permutations of three objects. Here we shall use the disjoint cycle notation for permutations so that, for example, (123) means the permutation which sends 1 to 2, 2 to 3, and 3 to 1.

The intermediate graph Y_3 corresponding to the subgroup $H = \{(1), (23)\}$ is also shown in the figure. An explanation of the method used to create these graphs is given in the following paragraphs. We then leave it to the reader to

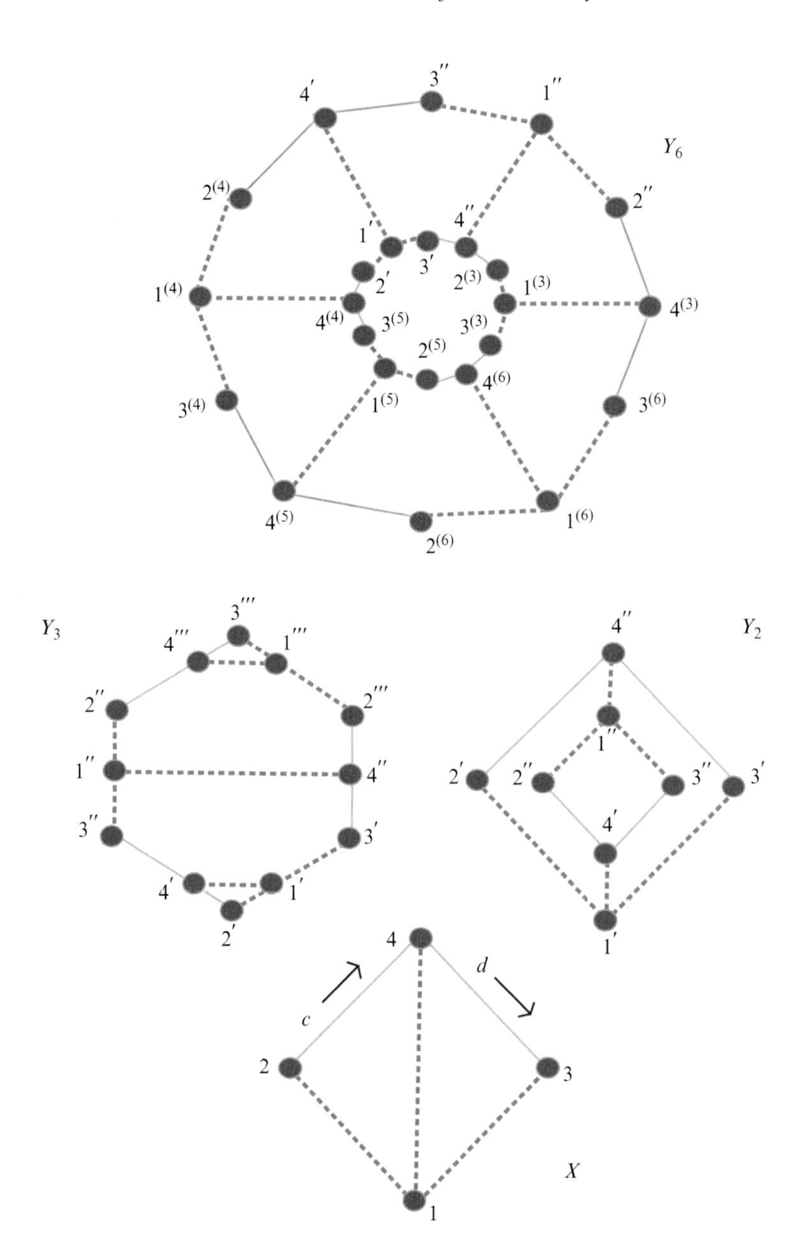

Figure 14.4 A six-sheeted normal cover Y_6 of X with a non-normal cubic intermediate cover Y_3 as well as a quadratic intermediate cover Y_2. Here the Galois group is the symmetric group $G = G(Y/X) = S_3$ and the subgroup $H = \{(1), (23)\}$ fixes Y_3. We write $a^{(1)} = (a, (1)), a^{(2)} = (a, (13)), a^{(3)} = (a, (132)), a^{(4)} = (a, (23)), a^{(5)} = (a, (123)), a^{(6)} = (a, (12))$, using standard cycle notation for elements of the symmetric group. A spanning tree in the base graph is shown by the dashed lines. The sheets of the covers are also shown in this way.

see how the intermediate graph Y_2 is created. Figure 14.4 should be compared with Figure 15.1.

The top graph Y_6 in Figure 14.4 is a normal 6-fold cover of the bottom graph X with Galois group S_3, the symmetric group of permutations of three objects. We make the identifications

$$a' = (a, (1)), \qquad a'' = (a, (13)), \qquad a^{(3)} = (a, (132)),$$
$$a^{(4)} = (a, (23)), \qquad a^{(5)} = (a, (123)), \qquad a^{(6)} = (a, (12)).$$

One way to construct this example is obtained by using a permutation representation of S_3. See Exercise 14.8 below. A spanning tree in the bottom graph X is shown by the dashed lines. The edges in X left out of the spanning tree (solid lines) generate the fundamental group of X. Let the directed edge from vertex 2 to vertex 4 be edge c and the directed edge from vertex 4 to vertex 3 be edge d.

We obtain the top graph Y_6 as follows. Take six copies of the spanning tree for X. This gives all the vertices of Y_6, but some edges are missing. Label as $x^{(i)}$ the vertex on the ith sheet of Y_6 projecting to vertex x in X. We lift the edge c to six edges in the top graph using the permutation $\sigma(c) = (14)(23)(56)$.[1] This permutation comes to us (not out of the blue but) from Exercise 14.8 below. It tells us to connect vertex $2^{(1)}$ with vertex $4^{(4)}$ in Y_6 and then connect vertices $2^{(4)}$ and $4^{(1)}$. After that, connect vertex $2^{(2)}$ with $4^{(3)}$ and vertex $2^{(3)}$ with $4^{(2)}$. Finally connect vertex $2^{(5)}$ with vertex $4^{(6)}$ and vertex $2^{(6)}$ with vertex $4^{(5)}$. Do a similar construction with $\sigma(d) = (12)(36)(45)$ to obtain the remaining six edges of Y_6. The permutations $\sigma(c)$ and $\sigma(d)$ have order 2 and they generate a subgroup of S_6 isomorphic to S_3. In Chapter 17 we will have more to say about this construction.

We can also identify S_3 with the dihedral group D_3 of rigid motions of an equilateral triangle. Let R be a 120° rotation of an equilateral triangle and F a flip. The we have $D_3 = \{I, R, R^2, F, FR, FR^2\}$, with $R^3 = I$ and $FR = R^2F$. We make the identifications

$$a' = (a, I), \qquad a'' = (a, FR^2), \qquad a^{(3)} = (a, R^2),$$
$$a^{(4)} = (a, FR), \qquad a^{(5)} = (a, R), \qquad a^{(6)} = (a, F)$$

and also $\sigma(c) = FR, \sigma(d) = FR^2$.

Next we want to construct the intermediate graph Y_3 corresponding to the subgroup $H = \{I, FR\}$. First we need the appropriate permutation

[1] Here $\sigma(c)$ denotes the normalized Frobenius automorphism corresponding to edge e, which we will define later; see Definition 16.1.

representation of G on the three cosets Hg_i, $i = 1, 2, 3$, with $g_1 = I, g_2 = FR^2, g_3 = F$. Since $\sigma(c) = FR$, we have $Hg_1\sigma(c) = Hg_1, Hg_2\sigma(c) = Hg_3$, and $Hg_3\sigma(c) = Hg_2$. Thus the cycle decomposition of the permutation corresponding to $\sigma(c)$ acting on cosets of H is (1)(23). Similarly the permutation of cosets of H corresponding to $\sigma(d) = FR^2$ is (12)(3).

We now construct the graph Y_3 intermediate to Y_6/X corresponding to H. First take three copies of the spanning tree in X. This gives all the vertices of Y_3, but some edges are missing. We label the sheets of Y_3 with single, double, and triple primes, respectively. The permutation (1)(23) tells us how to lift edge c in X to three edges in Y_3. We obtain edges in Y_3 from vertices $2'$ to $4'$, from $2''$ to $4'''$, and from $2'''$ to $4''$. Similarly the permutation (12)(3) tells us how to lift edge d in X to three edges in Y_3. We obtain the edges from $4'$ to $3''$, from $4''$ to $3'$, and from $4'''$ to $3'''$.

The three two-element subgroups of S_3 are all conjugate to H. Each will lead (by Theorem 14.5) to a graph isomorphic to Y_3, because we have not given the projections from Y_6 to Y_3. Without knowing these projections, the three conjugate cubic covers intermediate to Y_6/X are isomorphic, and the isomorphism preserves projections to X.

Exercise 14.7 Construct the intermediate graph Y_2 to Y_3/X in a similar way to that used to construct Y_3 above.

Exercise 14.8 List the elements of S_3, using cycle notation, as

$$g_1 = (1), \quad g_2 = (12), \quad g_3 = (123),$$

$$g_4 = (23), \quad g_5 = (132), \quad g_6 - (13).$$

Then write $g_i g = g_{\mu(g)}$, where $\mu(g) \in S_6$. Show that $\mu(23) = (14)(23)(56)$ and that $\mu(12) = (12)(36)(45)$.

Exercise 14.9 Create a covering of the cube graph which is normal and has as its Galois group a cyclic group of order 3.

We do not consider infinite graphs here, but it is possible to extend Galois theory to that situation. Then the Galois group of the universal covering tree T of a finite graph X can be identified with the fundamental group of X. This may make more sense after we have discussed Frobenius automorphisms in Chapter 16.

15

Behavior of primes in coverings

We seek analogs of the laws governing the behavior of prime ideals in extensions of algebraic number fields. Figure 1.5 shows what happens in a quadratic extension of the rationals. Figure 15.1 shows a non-normal cubic extension of the rationals. See Stark [118] for more information on these examples.

The graph theory analog of the example in Figure 15.1 is found in Figure 14.4 and Example 14.8. Figure 15.4 gives examples of primes that split in various ways in the non-normal cubic intermediate field.

So now let us consider the graph theory analog. The field extension is replaced by a graph covering Y/X, with projection map π. Suppose that $[D]$ is a prime in Y. Then $\pi(D)$ is a closed backtrackless tailless path in X, but it may not be primitive. There will, however, be a prime $[C]$ in X and an integer f such that $\pi(D) = C^f$. The integer f is independent of the choice of D in $[D]$.

Definition 15.1 If $[D]$ is a prime in a covering Y/X with projection map π and $\pi(D) = C^f$, where $[C]$ is a prime of X, we will say that $[D]$ is a **prime of Y above** $[C]$ or, more loosely, that D is a **prime above** C (written as $D|C$); $f = f(D, Y/X)$ is defined as the **residual degree** of D with respect to Y/X.

If Y/X is normal, for a prime C of X and a given integer j either every lift of C^j is closed in Y or no lift is closed. Thus the residual degree of $[D]$ above C is the same for all $[D]$ above C. This will not always be the case for non-normal extensions.

Definition 15.2 Let $g = g(D, Y/X)$ be the **number of primes** $[D]$ above $[C]$.

Since our covers are unramified, the analog of the ramification index is $e = e(D, Y/X) = 1$, and we will be able to prove the familiar formula from

field	ring	prime ideal	finite field
$K = F(e^{2\pi i/3})$	O_K	$\mathfrak{P}$	$O_K/\mathfrak{P}$
3 |	|	|	|
$F = \mathbb{Q}(\sqrt[3]{2})$	$O_F = \mathbb{Z}(\sqrt[3]{2})$	$\mathfrak{p}$	$O_F/\mathfrak{p}$
2 |	|	|	|
$\mathbb{Q}$	$\mathbb{Z}$	$p\mathbb{Z}$	$\mathbb{Z}/p\mathbb{Z}$

$g(\mathfrak{P}/\mathfrak{p}) =$ **# of such** $\mathfrak{P}$, $f(\mathfrak{P}/\mathfrak{p}) =$ **degree** $(O_K/\mathfrak{P} : O_F/\mathfrak{p})$

More details in Stark's article in From Number Theory to Physics, edited by Waldschmidt et al.

Splitting of rational primes in O_F

Type 1 $pO_F = \mathfrak{p}_1\mathfrak{p}_2\mathfrak{p}_3$, **with distinct $\mathfrak{p}_i$ of degree** 1 (**$p=31$ is 1st example), Frobenius of prime $\mathfrak{P}$ above $\mathfrak{p}_i$ has order** 1

density 1/6 **by Chebotarev**

Type 2 $pO_F = \mathfrak{p}_1\mathfrak{p}_2$, **with $\mathfrak{p}_1$ of degree** 1, **$\mathfrak{p}_2$ of degree** 2 (**$p = 5$ is 1st example), Frobenius of prime $\mathfrak{P}$ above $\mathfrak{p}_i$ has order** 2

density 1/2 **by Chebotarev**

Type 3 $pO_F = \mathfrak{p}$, **with $\mathfrak{p}$ of degree** 3 (**$p = 7$ is 1st example), Frobenius of $\mathfrak{P}$ above $\mathfrak{p}$ has order** 3

density 1/3 **by Chebotarev**

Figure 15.1 Example of the splitting of unramified primes in a non-normal cubic extension F of the rationals.

algebraic number theory for normal covers:

$$efg = d = \text{number of sheets of the cover.} \tag{15.1}$$

See part (6) of Proposition 16.5.

Example 15.3: Primes in the cube Y over primes in the tetrahedron X In Figure 15.2 we show a prime $[C]$ of length 3 in the tetrahedron X defined by $C = \langle a, d, c, a\rangle$. Here we list the vertices through which the path passes within $\langle\ \rangle$. The prime $[D]$ in the cube Y, with $D = \langle a', d'', c', a'', d', c'', a'\rangle$, has length 6 and is over $[C]$ in X. Let the Galois group be $G = G(Y/X) = \{1, \sigma\}$.

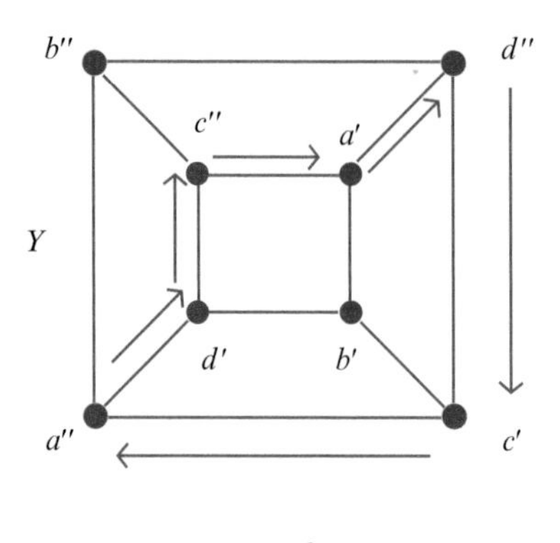

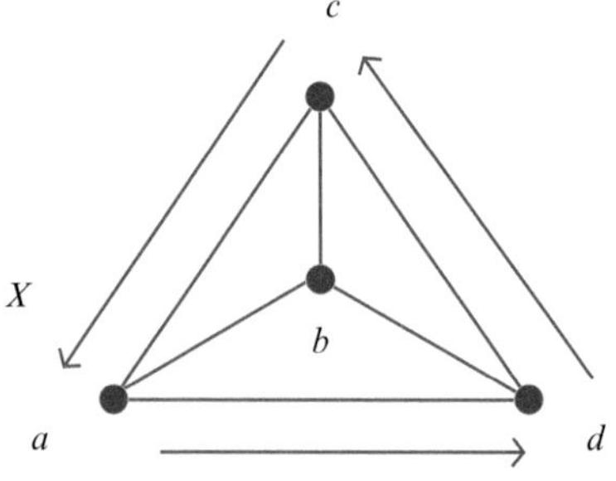

Figure 15.2 The splitting of prime C (see Example 15.3) with $f = 2$, $g = 1, e = 1$. There is one prime cycle D above C, and D is the lift of C^2.

We are using the notation $x' = (x, 1)$ and $x'' = (x, \sigma)$ in Y, for $x \in X$. Then $D = C_1\ (\sigma \circ C_1)$, where $C_1 = \langle a', d'', c', a'' \rangle$. Here $\nu(D) = 2\nu(C) = 6$. In this example $f = 2$ and $g = 1$.

A second example in Y/X is shown in Figure 15.3. In this case, the prime $[D]$ of Y is represented by $D = \langle a'', c', d'', b'', a'' \rangle$. Then D is a prime above C, where the prime $[C]$ is represented by $C = \langle a, c, d, b, a \rangle$ in X. Here $\nu(D) = \nu(C) = 4$, $f = 1$, and $g = 2$, since there is another prime D' in Y over C, also shown in Figure 15.3.

Example 15.4: Primes in a non-normal cubic cover Y_3, of $X = K_4 - e$, pictured in Figure 14.4 Figure 15.4 below gives examples of primes in X with various sorts of splittings in the non-normal cubic cover Y_3 of $X = K_4 - e$ from Example 14.8. The densities of the primes in various classes come from the Chebotarev density theorem, explained in Chapter 22 below.

Exercise 15.1 Figure 15.4 shows the splitting of primes in a non-normal cubic cover of $K_4 - e$. The prime in class 1 has length 10. Is it a prime of minimal length for $f = 1$ and $g = 3$?

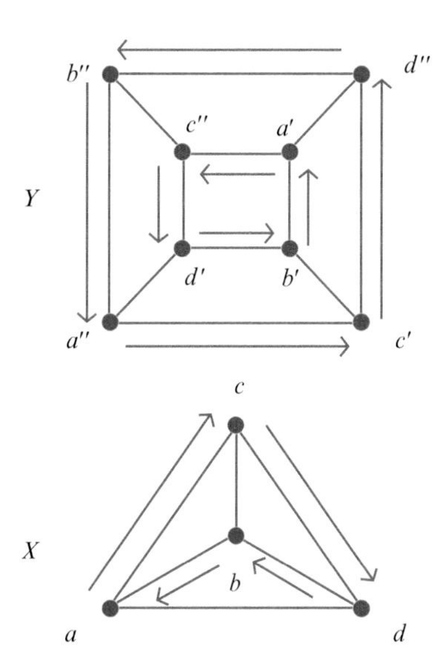

Figure 15.3 Picture of a prime C (see Example 15.3) which splits completely, i.e., $f = 1, g = 2, e = 1$. There are two prime cycles D, D' in the cube above C, each with the same length as C below in the tetrahedron.

Definition 15.5 If Y/X is normal and $[D]$ is a prime of Y over $[C]$ in X and σ is in $G(Y/X)$, we refer to $[\sigma \circ D]$ as a **conjugate prime** of Y over C.

We then have $f(\sigma \circ D, Y/X) = f(D, Y/X)$. If $f = f(D, Y/X)$ then, as σ runs through $G(Y/X)$, $\sigma \circ D$ runs through all possible lifts of C^f from X to Y, and thus the conjugates of $[D]$ account for all the primes of Y above $[C]$. That is, there are $d = |G(Y/X)|$ lifts of C^f starting on different sheets of Y, but only g of these lifts give rise to different primes of Y.

Exercise 15.2 Show that when the cover Y/X is not normal, formula (15.1) becomes

$$\sum_{i=1}^{g} f_i = d,$$

where d is the number of sheets of the cover and the f_i denote the residual degrees of the primes of Y above some fixed prime of X.

Given Y/X a (finite unramified) graph covering, suppose that $\widetilde{X}$ is intermediate to Y/X. Suppose that $\pi_1 : \widetilde{X} \longrightarrow X$, $\pi_2 : Y \longrightarrow \widetilde{X}$, and $\pi : Y \longrightarrow X$ are the covering maps, with $\pi = \pi_1 \circ \pi_2$. Let E be a prime of Y over the prime C of X and let $\pi_2(E) = D^{f_2}$, where D is a prime of $\widetilde{X}$ and $f_2 = f(E, Y/\widetilde{X})$.

Splitting of primes in non-normal cubic cover

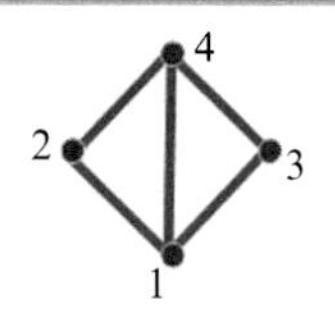

✵ **Class C1** **Path in X, vertices 14312412431.**

$f=1$, $g=3$; **three lifts to Y_3:**

1'4'3'''1'''2'''4''1''2''4'''3'1'

1''4''3''1''2''4'''1'''2'''4''3''1''

1'''4'''3'1'2'4'1'2'4'3'''1'''

Trivial Frobenius ⇒ density 1/6

✵ **Class C2** **Path in X, vertices 1241.**

Two lifts to Y_3:

1'2'4'1' $f=1$

1''2''4'''1'''2'''4''1'' $f=2$

Frobenius order 2 ⇒ density 1/2

✵ **Class C3** **Path in X, vertices 12431.**

$f=3$; **one lift to Y_3:**

1'2'4'3''1''2'''4''3''1''2''4'''3'1'

Frobenius order 3 ⇒ density 1/3

Figure 15.4 Splitting of primes in the non-normal cubic cover Y_3 of $K_4 - e$ pictured in Figure 14.4. This should be compared with Figure 15.1, which shows the splitting of primes in a non-normal cubic extension of the rational numbers.

Then we have the **transitivity property**

$$f(E, Y/X) = f(E, Y/\widetilde{X}) f(D, \widetilde{X}/X). \tag{15.2}$$

This is the graph theoretic analog of a result about the behavior of residual degrees of primes in extensions of algebraic number fields.

Exercise 15.3 Prove formula (15.2).
Hint: Note that $\pi\ (E) = C^{f(E,Y/X)}$ and $\pi_2(E) = D^{f(E,Y/\widetilde{X})}$.

16
Frobenius automorphisms

Before defining Artin L-functions of normal graph coverings, we should perhaps recall what Artin L-functions are and what they do for algebraic number fields. References for Artin L-functions of number fields are Lang [73] and Stark [116]. Figures 16.1 and 16.2 summarize some of the facts.

We want to find an analog of the Frobenius automorphism in number theory. References for the number theory version are Lang [73] and Stark [118].

Assume that Y is a normal cover of the graph X with Galois group G. We want to define the Frobenius automorphism $[Y/X, [D]]$ for a prime $[D]$ in Y over the prime $[C]$ in X. First we can define the normalized Frobenius automorphism $\sigma(p) \in G = G(Y/X)$ associated with a directed path p of X – the existence of which simplifies the graph theory version of things. This normalized Frobenius automorphism should be compared with the voltage assignment map in Gross and Tucker [47]. See Figure 16.3 for a summary of our definitions.

Definition 16.1 Suppose that Y/X is normal with Galois group $G = \mathrm{Gal}(Y/X)$. For a path p of X, Proposition 13.3 says there is a unique lift to a path $\tilde{p}$ of Y, starting on sheet 1, having the same length as p. If $\tilde{p}$ has its terminal vertex on the sheet labeled $g \in G$, define the **normalized Frobenius automorphism** $\sigma(p) \in G$ by $\sigma(p) = g$.

Exercise 16.1 Compute the normalized Frobenius automorphism $\sigma(C)$ for the paths C in the tetrahedron K_4 pictured in Figures 15.2 and 15.3.

Exercise 16.2 Compute the normalized Frobenius automorphism $\sigma(C)$ for the paths C in $K_4 - e$ pictured in Figure 15.4, when lifted to the top graph Y_6 in Figure 14.4.

$K \supset F$	**number fields with K/F Galois**
$O_K \supset O_F$	**rings of integers**
$\mathfrak{P} \supset \mathfrak{p}$	**prime ideals ($\mathfrak{p}$ unramified, i.e., $\mathfrak{p} \not\subset \mathfrak{P}^2$)**

Frobenius automorphism when $\mathfrak{p}$ is unramified:

$$\left(\frac{K/F}{\mathfrak{P}}\right) = \sigma_{\mathfrak{P}} \in \mathrm{Gal}(K/F), \quad \sigma_{\mathfrak{P}}(x) \equiv x^{N\mathfrak{p}} \pmod{\mathfrak{P}}, \quad x \in O_K, \quad N\mathfrak{p} = |O_K/\mathfrak{p}|$$

$\sigma_{\mathfrak{P}}$ **generates finite Galois group,** $\mathrm{Gal}((O_K/\mathfrak{P})/(O_F/\mathfrak{p}))$

determined by $\mathfrak{p}$ up to conjugation if $\mathfrak{P}/\mathfrak{p}$ unramified

$f(\mathfrak{P}/\mathfrak{p})$ = **order of** $\sigma_{\mathfrak{P}} = [O_K/\mathfrak{P} : O_F/\mathfrak{p}]$

$g(\mathfrak{P}/\mathfrak{p})$ = **number of primes of K dividing $\mathfrak{p}$**

Artin L-function for $s \in \mathbb{C}$, π a representation of $\mathrm{Gal}(K/F)$.
The formula is given only for unramified primes $\mathfrak{p}$ of F.
Pick $\mathfrak{P}$ a prime in O_K dividing $\mathfrak{p}$. Then

$$L(s,\pi) \text{ “=” } \prod_{\mathfrak{p}} \det\left(1 - \pi\left(\frac{K/F}{\mathfrak{P}}\right) N\mathfrak{p}^{-s}\right)^{-1}$$

where product is over primes $\mathfrak{p}$ of F.

Figure 16.1 Definition of the Frobenius symbol and Artin L-function of a Galois extension of number fields.

Lemma 16.2

(1) Suppose that p_1 and p_2 are two paths on X such that the terminal vertex of p_1 is the initial vertex of p_2. Then $\sigma(p_1 p_2) = \sigma(p_1)\sigma(p_2)$.

(2) If a path $p = e_1 \cdots e_s$, for directed edges $e_1, \ldots, e_s$ then we have $\sigma(p) = \sigma(e_1) \cdots \sigma(e_s)$.

Proof Part (1) If p_1 goes from a to b in X and p_2 goes from b to c in X then the lift $\widetilde{p_1}$ of p_1 starting on sheet 1 of Y goes from $(a, 1)$ to $(b, \sigma(p_1))$ and the lift $\widetilde{p_2}$ of p_2 starting on sheet 1 of Y goes from $(b, 1)$ to $(c, \sigma(p_2))$. See Figure 16.4. Therefore the lift of p_2 starting on sheet $\sigma(p_1)$ goes from $(b, \sigma(p_1))$ to $(c, \sigma(p_1)\sigma(p_2))$. This implies that the lift of $p_1 p_2$ starting on sheet 1 of Y will end on sheet $\sigma(p_1)\sigma(p_2)$.

Part (2) This follows easily from part (1). □

Now we can define Frobenius automorphisms and decomposition groups. See Figure 16.3 again.

Applications of Artin L-functions

❋ **Factorization** $\zeta_K(s) = \prod_{\pi} L(s,\pi)^{d_\pi}$

irreducible
degree d_π

❋ **Chebotarev density theorem**

$\forall\ \sigma$ in Gal(K/F), $\exists$ infinitely many prime ideals $\mathfrak{p}$ of O_F such that $\exists\ \mathfrak{P}$ in O_K dividing $\mathfrak{p}$ with Frobenius

$$\left(\frac{K/F}{\mathfrak{P}}\right) = \sigma$$

❋ **Artin conjecture** $L(s,\pi)$ **entire for non-trivial irreducible rep.** π (proved in fn fld case, not # fld case)

❋ **Stark conjectures For π not containing trivial rep.,**

$$\Longrightarrow \quad \lim_{s\to 0} s^a L(s,\pi) = \Theta(\pi) * R(\pi),$$

i.e., algebraic number × determinant of $a \times a$ matrix in linear forms with alg. coeffs. of logs of units of K and its conjugate fields/$\mathbb{Q}$. (First-order-0 case for fn fields probably done but not # fld case – Tate, Deligne, Hayes, Popescu.)

References

Stark's paper in **From Number Theory to Physics**, edited by Waldschmidt et al.
Stark, **Advances in Math.**, 17 (1975), 60–92
Lang or Neukirch, **Algebraic Number Theory**
Rosen, **Number Theory in Function Fields**

Figure 16.2 Applications of Artin L-functions of number fields.

Definition 16.3 Assume that Y/X is normal with Galois group G. Let $[C]$ be a prime on X such that C starts and ends at vertex a. Let $[D]$ be a prime of Y over C such that D starts and ends at vertex (a, g) on sheet $g \in G$ of Y. If the residual degree of D over C is f then D is the lifting of C^f which begins on sheet g. Suppose C itself lifts to a path $\widetilde{C}$ on Y starting on sheet g at (a, g) and ending on sheet h at (a, h). Define the **Frobenius automorphism** to be

$$[Y/X, D] = \left(\frac{Y/X}{D}\right) = hg^{-1} \in G.$$

Frobenius automorphism

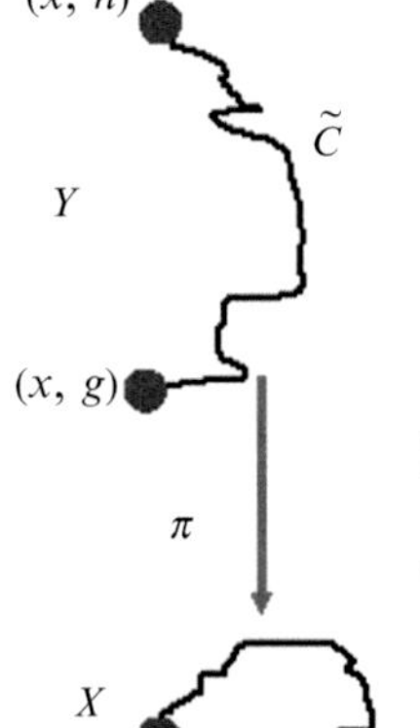

Suppose Y/X is a normal cover. Given a path p of X, there is a unique lift of p to a path $\tilde{p}$ on Y starting on sheet 1 and having the same length as p.

Here you may take p to be an edge if you like.

D a prime in Y above C a prime of X consists of f lifts of C.

$$\text{Frob}(D) = \left(\frac{Y/X}{D}\right) = hg^{-1} \in G = \text{Gal}(Y/X)$$

where hg^{-1} maps sheet g to sheet h

Normalized Frobenius: start lift of C on sheet 1, and end on sheet h: $\sigma(C)=h$.

Figure 16.3 Frobenius automorphism and the normalized version.

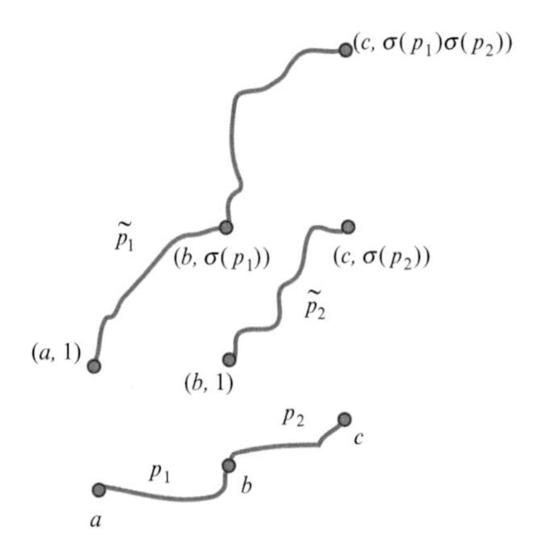

Figure 16.4 The map σ preserves the composition of paths.

Note that the Frobenius automorphism $[Y/X, D] = hg^{-1}$ maps sheet g of Y to sheet h of Y. To get the normalized version of the Frobenius, you have to take $g = 1$, the identity of G.

Our next definition yields a group analogous to one from algebraic number theory. The letter Z chosen for it corresponds to the German version of the name.

Definition 16.4 The **decomposition group** of D with respect to Y/X is

$$Z(D) = Z(D, Y/X) = \{\tau \in G \mid [\tau \circ D] = [D]\}.$$

The next proposition gives analogs of the usual properties of the Frobenius automorphism of a normal extension of number fields (as in Lang [73]).

Proposition 16.5 (Properties of the Frobenius automorphism) *As usual, Y/X is a normal d-sheeted covering with Galois group G.*

(1) For a prime cycle D in Y over C in X, the Frobenius automorphism is independent of the choice of D in its equivalence class $[D]$. Thus we can define $[Y/X, [D]] = [Y/X, D]$ without ambiguity.
(2) The order of $[Y/X, D]$ in G is the residual degree $f = f(D, Y/X)$.
(3) If $\tau \in G$ then $[Y/X, \tau \circ D] = \tau [Y/X, D] \tau^{-1}$.
(4) If D begins on sheet 1 then $[Y/X, D] = \sigma(C)$, the normalized Frobenius automorphism of Definition 16.1.
(5) The decomposition group $Z(D)$ is the cyclic subgroup of G of order f generated by $[Y/X, D]$. In particular, $Z(D)$ does not depend on the choice of D in its equivalence class $[D]$.
(6) For a prime cycle D in Y over C in X, if $f = f(D, Y/X)$ is as in Definition 15.1 and $g = g(D, Y/X)$ is as in Definition 15.2 then $d = fg$. (Here the ramification e *is assumed to be* 1.*)*

Proof Part (4) is proved by noting that the definitions of $\sigma(C)$ and $[Y/X, D]$ are then the same.

Parts (1) and (3) Suppose that C has initial (and terminal) vertex a in X and D is the lift of C^f beginning at vertex (a, μ_0) on sheet μ_0. In lifting C^f, we lift C a total of f times consecutively, beginning at (a, μ_0) and ending respectively at $(a, \mu_1), (a, \mu_2), \ldots, (a, \mu_{f-1}), (a, \mu_f)$, where $\mu_f = \mu_0$ and $\mu_j \neq \mu_0$ for $j = 1, 2, \ldots, f-1$. See Figure 16.5.

Suppose that (b, κ) is another vertex on D, where b is on C. Thus (b, κ) lies on one of the f consecutive lifts of C in Figure 16.5, say the rth. Vertex b splits C into two paths $C = p_1 p_2$, where b is the ending vertex of p_1 and the starting vertex of p_2. The vertex (b, κ) on Y is the ending vertex of the lift of p_1 to D starting at (a, μ_{r-1}). The lift of the version of C in $[C]$ starting at b, namely $p_2 p_1$, to a path on Y which starts at (b, κ) then ends at a vertex (b, λ) on D which lies on the $(r+1)$th consecutive lift of C.

Let $\widetilde{C}'$ be a path on Y from $(a, 1)$ to (a, μ_0) and let C' be the projection of $\widetilde{C}'$ to X. The vertices $(a, \mu_0), (a, \mu_1), (b, \kappa)$, and (b, λ) of Y are then the endpoints of the lifts of the paths

$$C', \quad C'C, \quad C'C^{r-1}p_1, \quad C'C^r p_1,$$

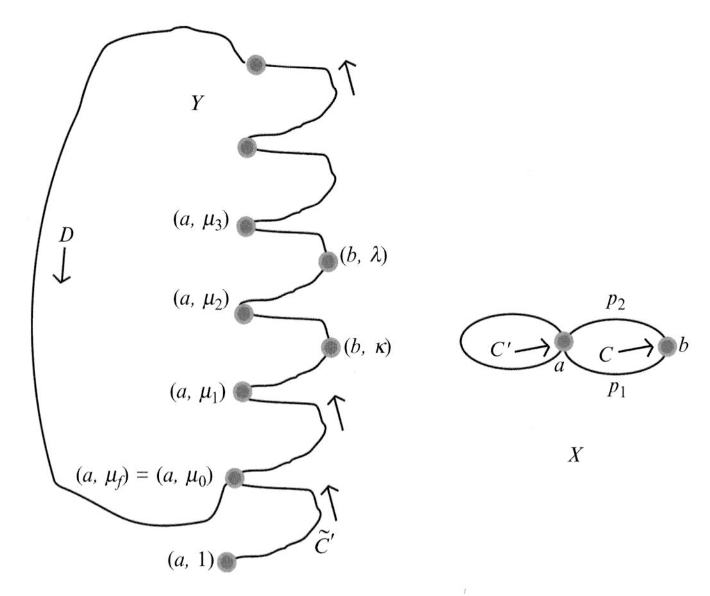

Figure 16.5 Part of the proof of Proposition 16.5. The vertex (b, κ) lies on the rth consecutive lift of C (shown for $r = 2$). The lift to a path in Y starting at (b, κ) of the version of C in $[C]$ starting at b ends at a vertex (b, λ) on the $(r + 1)$th consecutive lift of C.

respectively, to paths on Y starting at $(a, 1)$. Therefore, by Lemma 16.2, we have:

$$\mu_0 = \sigma(C'), \quad \mu_1 = \sigma(C'C) = \sigma(C')\sigma(C);$$
$$\kappa = \sigma(C'C^{r-1}p_1) = \sigma(C')\sigma(C)^{r-1}\sigma(p_1);$$
$$\lambda = \sigma(C'C^r p_1) = \sigma(C')\sigma(C)^r\sigma(p_1).$$

It follows that $[Y/X, D]$ is given by

$$\lambda\kappa^{-1} = \mu_1\mu_0^{-1} = \sigma(C')\sigma(C)\sigma(C')^{-1}.$$

This proves (1). It also proves (3) in the case $\tau = \mu_0^{-1} = \sigma(C')^{-1}$, and this suffices to prove (3) in general.

Part (2) As above, we see that, for each j,

$$\mu_j = \sigma(C'C^j) = \sigma(C')\sigma(C)^j$$

and thus

$$\mu_j\mu_0^{-1} = \sigma(C')\sigma(C)^j\sigma(C')^{-1} = [Y/X, D]^j. \tag{16.1}$$

This proves (2).

Part (5) Recall that $\tau \in Z(D)$ means that $\tau \circ D$ is equivalent to D in the sense of the equivalence relation with equivalence class $[D]$ in (2.2).

Suppose that $\tau \circ D$ is equivalent to D. If the picture is as in Figure 16.5 then, since $\tau \circ D$ also starts at a vertex projecting under $\pi : Y \to X$ to a, we must have $\tau\mu_0 = \mu_j$ for one of the μ_j above. Thus, for some j, we have $\tau = \mu_j\mu_0^{-1} = [Y/X, D]^j$ by (16.1).

Conversely, any such τ has $[\tau \circ D] = [D]$.

To see that the order of the decomposition group $Z(D)$ is $f = f(D, Y/X)$, note that $1 = \mu_j\mu_0^{-1} = [Y/X, D]^j$ iff f divides j.

Part (6) The Galois group $G = G(Y/X)$ of order d acts transitively on the primes $[D]$ above $[C]$. The subgroup $Z(D)$ is the subgroup of G fixing $[D]$. Since $Z(D)$ has order f, it follows that the number g of distinct $[D]$ is $|G/Z(D)| = d/f$. □

It remains only to discuss the behavior of the Frobenius automorphism with respect to intermediate coverings.

Theorem 16.6 (More properties of the Frobenius automorphism)

(1) Suppose that $\widetilde{X}$ is a covering intermediate to Y/X and that it corresponds to the subgroup H of $G = G(Y/X)$. Let $[D]$ be an equivalence class of prime cycles in Y such that D lies above $\widetilde{C}$ in $\widetilde{X}$. Let $f = f(D, Y/X) = f_1 f_2$, where $f_2 = f(D, Y/\widetilde{X})$ and $f_1 = f(\widetilde{C}, \widetilde{X}/X)$. Then f_1 is the minimal power of $[Y/X, D]$ which lies in H, and we have

$$[Y/X, D]^{f_1} = [Y/\widetilde{X}, D]. \tag{16.2}$$

(2) If, further, $\widetilde{X}$ is normal over X then as an element of $H\backslash G$ we have

$$[\widetilde{X}/X, \widetilde{C}] = H[Y/X, D].$$

Proof Part (1) Let C be the prime of X below $\widetilde{C}$. The Frobenius automorphism $[Y/\widetilde{X}, D]$ is found by lifting $\widetilde{C}$ from $\widetilde{X}$ to Y. This is the same as lifting C^{f_1} from X to Y, and the analysis in the proof of Proposition 16.5 (equation (16.1) in particular) gives equation (16.2)). The fact that f_1 is the minimal power of $[Y/X, D]$ which lies in H follows from the fact that

$$Z(Y/\widetilde{X}, D) = Z(Y/X, D) \cap H,$$

which we know to be cyclic of order f_2. Therefore, since $[Y/X, D]$ is of order $f_1 f_2$, we see that $[Y/X, D]^j$ cannot be in H if $j < f_1$.

Part (2) Now let $\widetilde{X}$ be normal over X. View $\widetilde{X}$ as having vertices $(v, H\tau)$ for $v \in X, \tau \in G$. Let D start and end at (a, μ_0) in Y and let $\widetilde{C}$ start and end at $(a, H\mu_0)$ in $\widetilde{X}$. If C in X lifts to a path in Y starting at (a, μ_0) and ending at (a, μ_1) then C lifts to a path in $\widetilde{X}$ starting at $(a, H\mu_0)$ and ending at $(a, H\mu_1)$. Then the statement follows from Definition 16.3, the definition of the Frobenius automorphism. $\square$

17

How to construct intermediate coverings using the Frobenius automorphism

Let us now explain how to construct intermediate coverings. First recall Example 14.8 and Figure 14.4. The following lemmas and theorem give the general construction that we followed in Example 14.8. Moreover they allow us to turn the tables, start with the intermediate cover Y_3, and produce its minimal normal cover Z with Y intermediate to Z/X.

Lemma 17.1 *Suppose that Y/X is normal with Galois group G. Fix a spanning tree T of X. Let $e_1, \ldots, e_r$ be the non-tree edges of X (i.e., those corresponding to generators of the fundamental group) with directions assigned. The r normalized Frobenius automorphisms $\sigma(e_j)$, $j = 1, \ldots, r$, generate G.*

Proof Since $\sigma(t) = 1$ for all edges t on the tree of X, for any path p on X, $\sigma(p)$ is a product of the $\sigma(e_j)$ and their inverses by Lemma 16.2. The graph Y is connected. Thus we can get to every sheet of Y by lifting paths of X to paths starting on sheet 1 of Y. It follows that any $g \in G$ is a product $\sigma(e_j)$. □

Lemma 17.2 *Suppose that Y/X is normal with Galois group G and $\widetilde{X}$ is an intermediate graph corresponding to the subgroup H of G. Let $H_0 = \bigcap_{g \in G} gHg^{-1}$. Then $H_0 = \{1\}$ if and only if there are no intermediate graphs, other than Y, which are normal over X and intermediate between Y and $\widetilde{X}$.*

Proof A normal intermediate graph covering $\widetilde{X}$ would correspond to a normal subgroup of G (a subgroup which must be contained in H) and conversely. Any normal subgroup of G contained in H is also contained in every conjugate of H and hence is contained in H_0. Since H_0 is a normal subgroup of G, the result is proved. □

Lemma 17.3 *Suppose that $\widetilde{X}$ is a covering of X and that Y/X is a normal covering of X of minimal degree such that $\widetilde{X}$ is intermediate to Y/X. Let $G = G(Y/X)$ and $H = G(Y/\widetilde{X})$. Let $Hg_1, \ldots, Hg_n$ be the right cosets of H. We have a one-to-one group anti-homomorphism μ from G into the symmetric group S_n that is defined by setting $\mu(g)(i) = j$ if $Hg_i g = Hg_j$.*

Proof By Exercise 17.1 below, $\mu(g')\mu(g) = \mu(gg')$. The kernel of μ is the set of $g \in G$ such that $Hg'g = Hg'\ \forall\ g' \in G$. This means that $Hg'gg'^{-1} = H\ \forall\ g' \in G$, which is equivalent to $g \in g'^{-1}Hg'\ \forall\ g' \in G$. By Lemma 17.2, $g = 1$ and μ is one-to-one. □

Exercise 17.1 Check the fact that $\mu(g')\mu(g) = \mu(gg')$, which was used in the preceding proof.

Now we will put these three lemmas together.

Theorem 17.4 *Let the graphs Y, $\widetilde{X}$, X, the groups G, H, and the representation μ be as in Lemma 17.3. Let T be a fixed spanning tree of X. Suppose that e is one of the non-tree edges of X. Let $\sigma(e)$ be the corresponding normalized Frobenius automorphism of G. Suppose that v is the starting vertex of e and that v' is the terminal vertex of e. If $\mu = \mu(\sigma(e))$ is the permutation of $1, \ldots, n$ such that $\mu(i) = \mu(\sigma(e))(i) = j$ then the directed edge e lifts to an edge in $\widetilde{X}$ starting at (v, Hg_i) and terminating at (v', Hg_j).*

Proof By the definition of μ, $Hg_i\sigma(e) = Hg_j$. This means that $g_i\sigma(e) = hg_j$ for some element $h \in H$. By the definition of $\sigma(e)$, the edge e lifts to an edge on Y from $(v, 1)$ to $(v', \sigma(e))$. If we apply g_i to this edge, we get an edge on Y starting at (v, g_i) and ending at $(v', g_i\sigma(e)) = (v', hg_j)$. Hence e lifts to a directed edge on $\widetilde{X}$ from (v, Hg_i) to (v', Hg_j). □

This theorem shows us how to create intermediate graphs given a normal cover, and it also allows us to construct the minimal normal cover Y of X having a given intermediate covering graph $\widetilde{X}$ of X as well as the Galois group $G(Y/X)$.

Example 17.5: Construction of the minimal normal cover Z of the cubic cover Y_3/X in Figure 14.4 Let G be the Galois group $G(Z/X)$ and H be the subgroup corresponding to Z. There are three cosets Hg_i, $i = 1, 2, 3$. Label the cosets so that Hg_1, Hg_2, Hg_3 correspond to the sheets of Y_3 labeled by single, double, and triple primes. According to Lemma 17.3, the permutation corresponding to $\sigma(c)$ is $(1)(23)$ and the permutation corresponding to $\sigma(d)$ is

$(12)(3)$. By Lemmas 17.1 and 17.3, these permutations generate an isomorphic copy of G in S_3. But the permutations (12) and (23) generate S_3. Thus $G = S_3$.

It follows that the minimal normal cover Z has six sheets. The vertices of Z are labeled (v, g) with $v \in V(X)$ and $g \in G$. We lift edge c by connecting $(2, g)$ with $(4, g\sigma(c))$. We lift edge d by connecting $(4, g)$ to $(4, g\sigma(d))$. The result is the graph $Z = Y_6$ pictured in Figure 14.4.

We cannot say which subgroup of S_3 corresponds to Y_3. We can only identify it up to conjugation. Why? We do not know which coset Hg_1, Hg_2, Hg_3 contains the identity. The three choices give three embeddings. Equivalently, we can relabel the sheets of Y_3 as the three cosets. On S_3 this relabeling is equivalent to a conjugation.

More examples of this theorem can be found in Figures 13.7, 13.8, and 13.9. We will give another series of examples based on the simple group of order 168 later.

Exercise 17.2 Check that the intermediate graphs in Figure 13.7 correspond, under the correspondence of the fundamental theorem of Galois theory, to the intermediate subgroups of $\mathbb{Z}_2 \times \mathbb{Z}_2$.

Exercise 17.3 Construct your own examples of n-fold cyclic covers of the graph $K_4 - e$ with all possible intermediate covers.

18
Artin L-functions

18.1 Brief survey on representations of finite groups

Artin L-functions involve representations of the Galois group. Thus the reader needs to know a bit about representations of finite groups. If this subject is new to you, perhaps it is best to restrict yourself to abelian or even cyclic groups at first.

Let $\mathbb{T}$ denote the multiplicative group of complex numbers of norm 1. Write $\mathbb{Z}_n = \mathbb{Z}/n\mathbb{Z}$ to denote the additive (cyclic) group of integers modulo n. A **representation of the cyclic group** $\pi : \mathbb{Z}_n \to \mathbb{T}$ is a group homomorphism. There are n distinct (inequivalent) representations of $\mathbb{Z}_n$ given by $\chi_a(x \bmod n) = \exp(2\pi i\, ax/n)$. Note that χ_a is well defined and does indeed change addition $\bmod\, n$ to multiplication of complex numbers on the unit circle. These are the functions on $\mathbb{Z}_n$ that can be used to find the **Fourier transform** $\hat{f}$ of any f: $\mathbb{Z}_n \to \mathbb{C}$ by writing

$$\hat{f}(a) = \sum_{x \in G} f(x)\overline{\chi_a(x)}. \tag{18.1}$$

This finite Fourier transform has analogous properties to the usual one on the real line and may be used to approximate the "real" version. In particular, one has **Fourier inversion**:

$$nf(-x) = \widehat{\widehat{f(x)}}\ \forall\, x \in \mathbb{Z}_n.$$

More generally, let G be any finite group. What is an irreducible unitary representation π of G? First, a **unitary representation** π is a group homomorphism $\pi : G \to U(n, \mathbb{C})$, where $U(n, \mathbb{C})$ is the group of $n \times n$ unitary complex matrices under matrix multiplication. A unitary matrix U means ${}^t\overline{U}U = I$. That the representation π is **irreducible** means we cannot block-upper-triangularize the matrices $\pi(g)$ by a uniform change of basis. The **degree**

of the representation π, denoted $d_\pi = n$, is the size of the matrices $\pi(g)$. If the group is abelian, the only irreducible representations are those of degree 1.

Two representations π_1 and π_2 of G are called **equivalent** (denoted $\cong$) if there is an invertible $n \times n$ complex matrix T such that $T\pi_1(g)T^{-1} = \pi_2(g)$ for all $g \in G$. Then $\widehat{G}$ is the set of all unitary irreducible inequivalent representations of G.

The **trivial representation**, denoted by 1, sends every element of G to the 1×1 matrix 1.

Given two representations π and ρ, we form the **direct sum** $\pi \oplus \rho$ by creating the block matrix

$$(\pi \oplus \rho)(g) = \begin{pmatrix} \pi(g) & 0 \\ 0 & \rho(g) \end{pmatrix}.$$

Every representation ρ of G is equivalent to a direct sum of irreducible representations:

$$\rho \cong \sum_{\pi \in \widehat{G}}^{\oplus} m_\pi \pi, \qquad \text{where the multiplicity } m_\pi \text{ is a non-negative integer.} \tag{18.2}$$

If the group is not abelian, we can find many examples of representations (though not necessarily irreducible ones) by a construction called induction. Suppose that H is a subgroup of G and ρ is a representation of H. Think of $\rho(g)$ as a linear map $\rho(g) : W \to W$, where W is a vector space over $\mathbb{C}$. Create the **representation induced by** ρ, denoted $\pi = \mathrm{Ind}_H^G \rho$, as follows. Define the vector space V to consist of functions on G which transform upon left action by H according to ρ. Then let $\pi(g)$ act on functions in V by right translation. That is, define $\pi = \mathrm{Ind}_H^G \rho$ by

$$V = \left\{ f : G \to W \,|\, f(hg) = \rho(h) f(g), \forall\, h \in H, \forall\, g \in G \right\},$$

$$(\pi(g) f)(x) = f(xg), \qquad \forall\, x, g \in G.$$

One then has the important theorem below. We call the representation $\mathrm{Ind}_{\{e\}}^G 1$ the **right regular representation**. Our theorem below says that the representation $\mathrm{Ind}_{\{e\}}^G 1$ (induced from the trivial representation on the trivial subgroup) is the mother of everything in $\widehat{G}$.

Theorem 18.1 *If e denotes the identity element of G and 1 is the trivial representation then*

$$\pi = \mathrm{Ind}_{\{e\}}^G 1 \cong \sum_{\pi \in \widehat{G}}^{\oplus} d_\pi \pi.$$

The notation means that the right regular representation is equivalent to that obtained from a block diagonal matrix built from all the inequivalent irreducible representations each taken a number of times equal to its degree.

Define the **character of the representation** ρ to be $\chi_\rho(g) = \mathrm{Tr}(\rho(g))$, where Tr stands for trace. For irreducible representations of abelian groups the characters are the representations, and we have seen them in earlier parts of the book when we investigated the eigenvalues of the adjacency matrix for Cayley graphs such as the Paley graph. See Section 9.2. It turns out that **the character determines the representation up to equivalence**. That is,

$$\chi_\pi = \chi_{\pi'} \qquad \text{iff} \qquad \pi \cong \pi'.$$

Clearly characters are invariant on **conjugacy classes** $\{g\} = \{xgx^{-1} | x \in G\}$. Thus one can create **character tables** indexed by the conjugacy classes $\{g\}$ of G and the inequivalent irreducible unitary representations $\pi \in \widehat{G}$. See Example 18.5 below for the character table of S_3 and look at Terras [133] for more examples.

There is also an analog of the finite Fourier transform in formula (18.1) and an inversion formula for non-abelian groups. See [133]. There are many other references for the representation theory of finite groups but few emphasize the connections with Fourier analysis.

We will need the following result of Frobenius. The proof can be found in [133], Chapter 16.

Theorem 18.2 (Frobenius character formula) *Suppose that H is a subgroup of the finite group G. Let σ be a representation of H. Define $\widetilde{\chi_\sigma}(y) = \chi_\sigma(y)$ if $y \in H$ and $\widetilde{\chi_\sigma}(y) = 0$ if $y \notin H$. Then $\pi = \mathrm{Ind}_H^G \sigma$ has character*

$$\chi_\pi(g) = \frac{1}{|H|} \sum_{x \in G} \widetilde{\chi_\sigma}\left(xgx^{-1}\right).$$

The following inner product of functions on G is of great use in representation theory. It can be used to see how many copies of an irreducible representation π of G are contained in an arbitrary representation ρ of G (the **multiplicity** of π in ρ).

Definition 18.3 Suppose that $f, g : G \to \mathbb{C}$. Define the **inner product** $\langle f, g \rangle_G$ by

$$\langle f, g \rangle_G = \frac{1}{|G|} \sum_{x \in G} f(x)\overline{g(x)}.$$

One can show that the characters of representations $\pi, \pi' \in \widehat{G}$ satisfy the **orthogonality relations**

$$\langle \chi_\pi, \chi_{\pi'} \rangle_G = \begin{cases} 0 & \text{if } \chi_\pi \neq \chi_{\pi'}, \\ 1 & \text{if } \chi_\pi = \chi_{\pi'}. \end{cases} \tag{18.3}$$

The earlier formula (10.5) was a special case of these orthogonality relations.

It follows from these orthogonality relations that the **multiplicity** m_π of $\pi \in \widehat{G}$ in a representation ρ of G from formula (18.2) can be computed from

$$m_\pi = \langle \chi_\pi, \chi_\rho \rangle_G .$$

This explains how the character determines the representation up to equivalence.

We will find that a second theorem of Frobenius is very useful.

Theorem 18.4 (Frobenius reciprocity law) *Under the same hypotheses as the preceding theorem, with $\pi = \mathrm{Ind}_H^G \sigma$, we have, for any representation ρ of G,*

$$\langle \chi_\rho, \ \chi_\pi \rangle_G = \langle \chi_{\rho|_H}, \ \chi_\sigma \rangle_H .$$

Here $\rho|_H$ denotes the restriction of ρ to H.

The following example is a favorite for understanding the zeta functions of the covering in Figure 14.4.

Example 18.5: Character table for S_3 We want to consider the character table for the symmetric groups S_3 of permutations of three objects. As usual, we employ the disjoint cycle notation. Then the conjugacy classes are easily seen to be $\{(1)\}$, $\{(12)\}$, and $\{(123)\}$. There are two obvious one-dimensional representations: the trivial representation χ_1 and the representation $\chi_1'(\sigma) = (-1)^{\mathrm{sgn}\,\sigma}$, where sgn $\sigma =$ the number of transpositions needed if we write the permutation σ as a product of transpositions (ab).

The third element of $\widehat{G}$ is a degree-2 representation ρ, which can be obtained as follows. For any $\sigma \in G$, define the 3×3 matrix $M(\sigma)$ by

$$M(\sigma) \begin{pmatrix} v_1 \\ v_2 \\ v_3 \end{pmatrix} = \begin{pmatrix} v_{\sigma^{-1}(1)} \\ v_{\sigma^{-1}(2)} \\ v_{\sigma^{-1}(3)} \end{pmatrix} .$$

Table 18.1 *Character table for S_3*

$\widehat{G}\backslash\{g\}$	$\{(1)\}$	$\{(12)\}$	$\{(123)\}$
χ_1	1	1	1
$\chi_1^{'}$	1	-1	1
χ_ρ	2	0	-1

One sees that M is a representation of G which induces a degree-2 representation ρ on the subspace

$$W = \left\{ \begin{pmatrix} v_1 \\ v_2 \\ v_3 \end{pmatrix} \middle| \, v_1 + v_2 + v_3 = 0 \right\}.$$

The **character table** for S_3 is given in Table 18.1.

The following exercise is useful in gaining understanding of the factorization of zeta functions for normal S_3 covers.

Exercise 18.1 (a) Define $\pi_1 = \mathrm{Ind}_{\{(1)\}}^{S_3} 1$. Show that $\chi_{\pi_1} = \chi_1 + \chi_1^{'} + 2\chi_\pi$.
(b) Define $\pi_2 = \mathrm{Ind}_{\{(1),(23)\}}^{S_3} 1$. Show that $\chi_{\pi_2} = \chi_1 + \chi_\pi$.
(c) Define $\pi_3 = \mathrm{Ind}_{\{(1),(123),(132)\}}^{S_3} 1$. Show that $\chi_{\pi_3} = \chi_1 + \chi_1^{'}$.

18.2 Definition of the Artin–Ihara L-function

Suppose that Y is a normal unramified covering of X with Galois group $G = G(Y/X)$.

Definition 18.6 If ρ is a representation of G with degree $d = d_\rho$ and u is a complex variable with $|u|$ sufficiently small, define the **Artin–Ihara L-function**

$$L(u, \rho, Y/X) \equiv L(u, \rho) = \prod_{[C]} \det\left(I - \rho\left([Y/X, D]\right) u^{\nu(C)}\right)^{-1},$$

where the product runs over the primes $[C]$ of X and $[D]$ is arbitrarily chosen from the primes in Y above C. Here $[Y/X, D]$ is the Frobenius automorphism of Definition 16.3 and $\nu(C)$ is the length of a path C representing the prime $[C]$.

The Frobenius automorphism is unique only up to conjugacy, but this does not matter thanks to the determinant. When the representation ρ is trivial (i.e., it equals 1), the L-function becomes the Ihara zeta function of Definition 2.2:

$$L(u,1,Y/X)=\zeta_X(u). \tag{18.4}$$

As in the case of the Ihara zeta function, the Artin L-function is the reciprocal of a polynomial, owing to the existence of analogs of the two determinant formulas. The simplest such formula involves a W_1-type matrix.

Definition 18.7 Define the **Artinized edge adjacency matrix** $W_{1,\rho}$ to be a matrix built up out of blocks corresponding to directed edges e and f:

$$\left(W_{1,\rho}\right)_{e,f}=\rho(\sigma(e))\,(W_1)_{e,f}$$

Here $\sigma(e)$ is the normalized Frobenius automorphism, attached to directed edge e, from Definition 16.1 and W_1 is the edge adjacency matrix from Definition 4.1.

Theorem 18.8 *We have* $L(u,\rho,Y/X)^{-1}=\det\left(I-uW_{1,\rho}\right)$.

Proof We can imitate the proof of formula (4.4). Since $\exp \operatorname{Tr} A=\det\exp A$ for any matrix A (by Exercise 4.1), we have

$$\begin{aligned}
&\log L(u,\rho,Y/X)\\
&=\sum_{\substack{[P]\\ \text{prime}}}\log\det\left(I-\rho(\sigma(P))u^{\nu(P)}\right)=-\sum_{[P]}\operatorname{Tr}\log\left(I-\rho(\sigma(P))u^{\nu(P)}\right)\\
&=\operatorname{Tr}\left(\sum_{[P]}\sum_{j\geq 1}\frac{1}{j}\rho\left(\sigma(P)\right)^{j}u^{j\nu(P)}\right)=\operatorname{Tr}\left(\sum_{P}\sum_{j\geq 1}\frac{1}{\nu(P^j)}\rho\left(\sigma(P^j)\right)u^{\nu(P^j)}\right)\\
&=\sum_{\substack{C\\ \text{closed}\\ \text{no backtrack or tail}}}\frac{1}{\nu(C)}\chi_\rho\left(\sigma(C)\right)u^{\nu(C)}.
\end{aligned}$$

We have also used the fact that the equivalence class $[P]$ has $\nu(P)$ primitive paths, and we needed to know that the normalized Frobenius automorphism $\sigma(C)$ is multiplicative as well as the fact that any closed path C without backtracking and without tail has the form $C=P^j$ for some prime $[P]$ and some positive integer j. Next, set $B=W_{1,\rho}$ and note that

$$\begin{aligned}\text{Tr } B^m &= \text{Tr}\left(\sum_{e_1,\ldots,e_m} b_{e_1e_2}b_{e_2e_3}\cdots b_{e_{m-1}e_m}b_{e_me_1}\right)\\ &= \text{Tr}\left(\sum_{e_1,\ldots,e_m} \rho(\sigma(e_1))\,\rho(\sigma(e_2))\cdots\rho(\sigma(e_{m-1}))\,\rho(\sigma(e_m))\right)\\ &= \sum_{\substack{C,\ \nu(C)=m\\ \text{closed}\\ \text{no backtrack or tail}}} \text{Tr}(\rho(\sigma(C))).\end{aligned}$$

Note that the sum in the last line is over $e_1, \ldots, e_m$ such that $e_1 \cdots e_m$ is a closed path without backtrack or tail. Therefore using $\exp \text{Tr } A = \det \exp A$ from Exercise 4.1 again, we obtain

$$\begin{aligned}\log L(u, \rho, Y/X) &= \sum_{m\geq 1} \frac{1}{m} \sum_{\nu(C)=m} \chi_\rho(\sigma(C))u^m\\ &= \sum_{m\geq 1} \frac{1}{m} \sum_{\nu(C)=m} \text{Tr}(\rho(\sigma(C)))u^m\\ &= \text{Tr}\left(\sum_{m\geq 1} \frac{1}{m} W_{1,\rho}^m u^m\right) = \text{Tr}\left(\log(I - uW_{1,\rho})^{-1}\right)\\ &= \log\det(I - uW_{1,\rho}),\end{aligned}$$

and the proof is complete. □

Example 18.9: Klein 4-group cover of dumbbell and intermediate covers Recall Figure 13.7 and Example 13.12. First look at the three quadratic intermediate covers and their Artin L-functions. Scientific Workplace can be used on a PC to compute the 4×4 determinants. First we need the edge adjacency matrix W_1 for the base graph, i.e., the dumbbell X where the edges of X are ordered as $a, b, c, a^{-1}, b^{-1}, c^{-1}$:

$$W_1 = \begin{pmatrix} 1 & 0 & 1 & 0 & 0 & 0\\ 0 & 1 & 0 & 0 & 0 & 1\\ 0 & 1 & 0 & 0 & 1 & 0\\ 0 & 0 & 1 & 1 & 0 & 0\\ 0 & 0 & 0 & 0 & 1 & 1\\ 1 & 0 & 0 & 1 & 0 & 0 \end{pmatrix}.$$

The zeta function of the dumbbell is

$$\zeta_X(u)^{-1} = \det(I - uW_1) = \det\begin{pmatrix} 1-u & 0 & -u & 0 & 0 & 0 \\ 0 & 1-u & 0 & 0 & 0 & -u \\ 0 & -u & 1 & 0 & -u & 0 \\ 0 & 0 & -u & 1-u & 0 & 0 \\ 0 & 0 & 0 & 0 & 1-u & -u \\ -u & 0 & 0 & -u & 0 & 1 \end{pmatrix}$$

$$= -4u^6 + 8u^5 - 3u^4 - 4u^3 + 6u^2 - 4u + 1$$

$$= -(u+1)(u-1)^2(2u-1)\left(2u^2 - u + 1\right).$$

Now we need the normalized Frobenius automorphisms for the three intermediate quadratic covers Y', Y'', Y'''. Write the Galois group as $G = \mathbb{Z}_2 = \{0, 1 \pmod 2\}$. Then

$$Y': \quad \sigma(a) = 1 (\mathrm{mod}\, 2), \quad \sigma(b) = 1 (\mathrm{mod}\, 2), \quad \sigma(c) = 0 (\mathrm{mod}\, 2).$$

$$Y'': \quad \sigma(a) = 0 (\mathrm{mod}\, 2), \quad \sigma(b) = 1 (\mathrm{mod}\, 2), \quad \sigma(c) = 0 (\mathrm{mod}\, 2).$$

$$Y''': \quad \sigma(a) = 1 (\mathrm{mod}\, 2), \quad \sigma(b) = 0 (\mathrm{mod}\, 2), \quad \sigma(c) = 0 (\mathrm{mod}\, 2).$$

Let the 2 representations of G be denoted 1 for the trivial representation and ρ, where $\rho(0(\mathrm{mod}\, 2)) = 1$ and $\rho(1(\mathrm{mod}\, 2)) = -1$. Then the matrices W_σ are found below for each of the three intermediate quadratic covers.

For Y', we get the Artin L-function

$$L(u, \rho, Y'/X)^{-1} = \det(I - uW_{1,\rho}) = \det\begin{pmatrix} 1+u & 0 & u & 0 & 0 & 0 \\ 0 & 1+u & 0 & 0 & 0 & u \\ 0 & -u & 1 & 0 & -u & 0 \\ 0 & 0 & u & 1+u & 0 & 0 \\ 0 & 0 & 0 & 0 & 1+u & u \\ -u & 0 & 0 & -u & 0 & 1 \end{pmatrix}$$

$$= -4u^6 - 8u^5 - 3u^4 + 4u^3 + 6u^2 + 4u + 1$$

$$= -(u+1)^2(u-1)(2u+1)\left(2u^2 + u + 1\right).$$

For Y'', we get the Artin L-function

$$L(u,\rho,Y''/X)^{-1}=\det(I-uW_{1,\rho})=\det\begin{pmatrix}1-u&0&-u&0&0&0\\0&1+u&0&0&0&u\\0&-u&1&0&-u&0\\0&0&-u&1-u&0&0\\0&0&0&0&1+u&u\\-u&0&0&-u&0&1\end{pmatrix}$$

$$=-4u^6+5u^4-2u^2+1$$

$$=(u-1)(u+1)\left(-4u^4+u^2-1\right).$$

For Y''', we get the Artin L-function

$$L(u,\rho,Y'''/X)^{-1}=\det(I-uW_{1,\rho})=\det\begin{pmatrix}1+u&0&u&0&0&0\\0&1-u&0&0&0&-u\\0&-u&1&0&-u&0\\0&0&u&1+u&0&0\\0&0&0&0&1-u&-u\\-u&0&0&-u&0&1\end{pmatrix}$$

$$=-4u^6+5u^4-2u^2+1$$

$$=(u-1)(u+1)\left(-4u^4+u^2-1\right).$$

So we find that the zeta functions of the three intermediate quadratic covers are

$$\zeta_{Y'}(u)^{-1}=\zeta_X(u)^{-1}L(u,\rho,Y'/X)^{-1}$$

$$=(u+1)^3(u-1)^3(2u+1)(2u-1)\left(2u^2+u+1\right)\left(2u^2-u+1\right),$$

$$\zeta_{Y''}(u)^{-1}=\zeta_X(u)^{-1}L(u,\rho,Y''/X)^{-1}$$

$$=(u+1)^2(u-1)^3(2u-1)\left(2u^2-u+1\right)\left(-4u^4+u^2-1\right),$$

$$\zeta_{Y'''}(u)^{-1}=\zeta_X(u)^{-1}L(u,\rho,Y'''/X)^{-1}$$

$$=(u+1)^3(u-1)^2(2u+1)\left(2u^2+u+1\right)\left(-4u^4+u^2-1\right).$$

Now let us consider the top graph Z in Figure 13.7. The Galois group is $\mathbb{Z}_2 \times \mathbb{Z}_2 = \{(a, b) | a, b \in \mathbb{Z}_2\}$. The representations are χ_c, where $c = (c_1, c_2) \in \mathbb{Z}_2 \times \mathbb{Z}_2$ and, for $x = (x_1, x_2)$,

$$\chi_c(x) = e^{2\pi i(c_1x_1+c_2x_2)/2} = (-1)^{c_1x_1+c_2x_2}.$$

We identify sheet 1 (the lowest sheet in the figure) with the element (0, 0) of the Galois group. Then we identify sheet 2 (the next highest sheet) with element (0, 1), sheet 3 (the next highest sheet) with (1, 0) and sheet 4 (the top sheet) with (1, 1). The normalized Frobenius automorphisms are

$$\sigma(a) = (0, 1), \qquad \sigma(b) = (1, 1), \qquad \sigma(c) = (0, 0).$$

Then

$$\chi_{(0,1)}(\sigma(a)) = (-1)^1 = -1, \quad \chi_{(0,1)}(\sigma(b)) = (-1)^1 = -1,$$
$$\chi_{(0,1)}(\sigma(c)) = (-1)^0 = 1;$$
$$\chi_{(1,0)}(\sigma(a)) = (-1)^0 = 1, \quad \chi_{(1,0)}(\sigma(b)) = (-1)^1 = -1,$$
$$\chi_{(1,0)}(\sigma(c)) = (-1)^0 = 1;$$
$$\chi_{(1,1)}(\sigma(a)) = (-1)^1 = -1, \quad \chi_{(1,1)}(\sigma(b)) = (-1)^2 = 1,$$
$$\chi_{(1,1)}(\sigma(c)) = (-1)^0 = 1.$$

We know that

$$L(u, \chi_{(0,0)}, Z/X) = \zeta_X(u) = -(u+1)(u-1)^2(2u-1)\left(2u^2 - u + 1\right).$$

The three new Artin L-functions for Z/X from the non-trivial representations are found easily now. First note that $\chi_{(0,1)}(\sigma(e)) = \rho(\sigma(e))$ for the graph Y'. This means that

$$\begin{aligned} L(u, \chi_{(0,1)}, Z/X)^{-1} &= L(u, \rho, Y'/X)^{-1} \\ &= -(u-1)(u+1)^2(2u+1)\left(2u^2 + u + 1\right). \end{aligned}$$

Then we see that $\chi_{(1,0)}(\sigma(e)) = \rho(\sigma(e))$ for the graph Y'', implying that

$$L(u, \chi_{(1,0)}, Z/X)^{-1} = L(u, \rho, Y''/X)^{-1} = (u-1)(u+1)\left(-4u^4 + u^2 - 1\right).$$

Finally $\chi_{(1,1)}(\sigma(e)) = \rho(\sigma(e))$ for the graph Y''', implying that

$$L(u, \chi_{(1,1)}, Z/X)^{-1} = L(u, \rho, Y'''/X)^{-1} = (u-1)(u+1)\left(-4u^4 + u^2 - 1\right).$$

It is not surprising that the L-functions corresponding to the covers Y'' and Y''' are the same. These graphs are isomorphic as abstract graphs, while Y' is not isomorphic to Y''. The L-function for Y' comes from that for the trivial representation on replacing u by $-u$.

The reader can now check using the Ihara determinant formula that, as stated in the corollary to the proposition below, the zeta function of the top quartic cover Z is the product of all the Artin L-functions of the Galois group:

$$\begin{aligned}\zeta_Z(u) &= L(u, \chi_{(0,0)}, Z/X)L(u, \chi_{(0,1)}, Z/X) \\ &\quad \times L(u, \chi_{(1,0)}, Z/X)L(u, \chi_{(1,1)}, Z/X) \\ &= (u+1)^5(u-1)^5(2u+1)(2u-1) \\ &\quad \times \left(2u^2+u+1\right)\left(2u^2-u+1\right)\left(-4u^4+u^2-1\right)^2.\end{aligned}$$

Note that Z is obtained from an 8-cycle by replacing every other edge with a double edge. It follows that

$$\zeta_X(u)^2\zeta_Z(u) = \zeta_{Y'}(u)\zeta_{Y''}(u)\zeta_{Y'''}(u).$$

18.3 Properties of Artin–Ihara L-functions

Now we want to list the properties of Artin L-functions of normal graph coverings. They are essentially the same as those for the Artin L-functions of Galois extensions of number fields from Lang [73].

Proposition 18.10 (Properties of the Artin–Ihara L-function) *Assume that Y/X is a normal covering with Galois group G.*

(1) $L(u, \rho_1 \oplus \rho_2) = L(u, \rho_1)L(u, \rho_2)$.

(2) Suppose that $\widetilde{X}$ is intermediate to Y/X and assume that $\widetilde{X}/X$ is normal, $G = \mathrm{Gal}(Y/X)$, $H = \mathrm{Gal}(Y/\widetilde{X})$. Let ρ be a representation of $G/H \cong \mathrm{Gal}(\widetilde{X}/X)$. Thus ρ can be viewed as a representation of G (the lift of ρ). Then

$$L(u, \rho, Y/X) = L(u, \rho, \widetilde{X}/X).$$

(3) (Induction property) If $\widetilde{X}$ is a cover intermediate to the normal cover Y/X and ρ is a representation of $H = \mathrm{Gal}(Y/\widetilde{X})$ then let $\rho^\# = \mathrm{Ind}_H^G\,\rho$, that is, the representation induced by ρ from H up to G. Then

$$L(u, \rho^\#, Y/X) = L(u, \rho, Y/\widetilde{X}).$$

Here we do not assume that $\widetilde{X}$ is normal over X.

Proof Only part (3) really requires some effort. We will postpone the proofs until the next chapter when we consider the more general case of edge L-functions. □

From these properties and Theorem 18.1, we have the following corollary.

Corollary 18.11 (Factorization of the Ihara zeta function of an unramified normal extension of graphs) *Suppose that Y/X is normal with Galois group $G = G(Y/X)$. Let $\widehat{G}$ be a complete set of inequivalent irreducible unitary representations of G. Then*

$$\zeta_Y(u) = L(u, 1, Y/Y) = \prod_{\rho \in \widehat{G}} L(u, \rho, Y/X)^{d_\rho}.$$

Proof Take $Y = \widetilde{X}$ in part (3) of Proposition 18.10. The corresponding subgroup H of G is $H = \{e\}$. Then let $\rho = 1$, the trivial representation on H; then $\rho^{\#} = \mathrm{Ind}_{\{e\}}^{G} 1$ is the right regular representation. Theorem 18.1 says that $\rho^{\#} = \mathrm{Ind}_{\{e\}}^{G} 1 \cong \sum_{\pi \in \widehat{G}}^{\oplus} d_\pi \pi$. Use formula (18.4), as well as parts (1) and (3) of Proposition 18.10, to see that

$$\begin{aligned} \zeta_Y(u) &= L(u, 1, Y/Y) = L(u, \rho^{\#}, Y/X) \\ &= L\left(u, \sum_{\pi \in \widehat{G}}^{\oplus} d_\pi \pi, Y/X\right) = \prod_{\rho \in \widehat{G}} L(u, \rho, Y/X)^{d_\rho}. \end{aligned}$$

□

We now define some matrices associated with a representation ρ of $G(Y/X)$, where Y/X is a finite unramified normal covering of graphs.

Definition 18.12 For $\sigma, \tau \in G$ and vertices $a, b \in X$, define the matrix $A(\sigma, \tau)$ to be the $n \times n$ matrix given by setting the entry $A(\sigma, \tau)_{a,b}$ equal to the number of directed edges in Y from (a, σ) to (b, τ). Here every undirected edge of Y has been given both directions.

Except when (a, σ) and (b, τ) are the same vertex on Y (i.e., $a = b$ and $\sigma = \tau$), and even then if there is no loop at $(a, \sigma) = (b, \tau)$, $A(\sigma, \tau)_{a,b}$ is simply the number of undirected edges on Y connecting (a, σ) to (b, τ). However, if there is a loop at $(a, \sigma) = (b, \tau)$ then it is counted in both directions and thus the undirected loop is counted twice. It follows from the Exercise below that we can write

$$A(\sigma, \tau) = A\big(1, \sigma^{-1}\tau\big) \equiv A\big(\sigma^{-1}\tau\big). \tag{18.5}$$

Exercise 18.2 Show that

$$A(\sigma, \tau) = A(1, \sigma^{-1}\tau).$$

Definition 18.13 If ρ is a representation of $G(Y/X)$ and $A(\sigma, \tau)$ is given by Definition 18.12 and formula (18.5), define the **Artinized adjacency matrix** A_ρ by

$$A_\rho = \sum_{\sigma \in G} A(\sigma) \otimes \rho(\sigma).$$

Also set

$$Q_\rho = Q \otimes I_d,$$

where Q is the $|X| \times |X|$ diagonal matrix with diagonal entries corresponding to $a \in X$ and given by $q_a =$ (degree a) $- 1$, and d is the degree of ρ.

Theorem 18.14 (Block diagonalization of the adjacency matrix of a normal cover) *Suppose that Y/X is normal with Galois group $G = G(Y/X)$. Let $\widehat{G}$ be a complete set of inequivalent irreducible unitary representations of G. Then one can block-diagonalize the adjacency matrix of Y with diagonal blocks A_ρ, each listed d_ρ times as ρ runs through $\widehat{G}$.*

Proof The adjacency operator on Y may be viewed as coming from the representation $\mathrm{Ind}_{\{e\}}^G 1$ with the decomposition in Theorem 18.1. List the vertices of Y as (x, τ), $x \in X$, $\tau \in G$. This decomposes A_Y into $n \times n$ blocks, where $n = |X|$, with blocks $A(\sigma, \tau) = A(\sigma^{-1}\tau)$ given by Definition 18.12, using formula (18.5) for $\sigma, \tau \in G$. This means that $\sigma \in G$ is acting on the function $A : G \to \mathbb{R}$ via $\lambda(\sigma)A(\tau) = A(\sigma^{-1}\tau)$, with $\sigma, \tau \in G$. Then λ is the left regular representation of G. This is equivalent to $\mathrm{Ind}_{\{e\}}^G 1$. It follows from Theorem 18.1 that A_Y has block decomposition into blocks A_ρ corresponding to $\rho \in \widehat{G}$, each listed d_ρ times. □

Now we can generalize Theorem 2.5.

Theorem 18.15 (Ihara theorem for Artin L-functions) *With the hypotheses and definitions above, we have*

$$L(u, \rho, Y/X)^{-1} = \left(1 - u^2\right)^{(r-1)d} \det\left(I - A_\rho u + Q_\rho u^2\right).$$

Here r is the rank of the fundamental group of X.

Proof We will postpone the proof until the next chapter, where we give the L-function version of Bass's proof of Theorem 2.5. For this we will need edge Artin L-functions. □

18.4 Examples of factorizations of Artin–Ihara L-functions

Example 18.16: The cube over the tetrahedron See Figure 13.4, where the action of the group $G = G(Y/X) = \{1, \sigma\}$ on Y is denoted by primes, i.e., $x' = (x, 1)$ and $x'' = (x, \sigma)$, for $x \in X$. In this case the representations of G are the trivial representation $\rho_0 = 1$ and the representation ρ defined by $\rho(1) = 1$, $\rho(\sigma) = -1$. So $Q_\rho = 2I_4$. There are two cases.

Case 1 **The representation $\rho_0 = 1$.**
Here $A_1 = A(1) + A(\sigma) = A$, where

$$A(1) = \begin{pmatrix} 0 & 1 & 0 & 0 \\ 1 & 0 & 1 & 1 \\ 0 & 1 & 0 & 0 \\ 0 & 1 & 0 & 0 \end{pmatrix}, \qquad A(\sigma) = \begin{pmatrix} 0 & 0 & 1 & 1 \\ 0 & 0 & 0 & 0 \\ 1 & 0 & 0 & 1 \\ 1 & 0 & 1 & 0 \end{pmatrix},$$

and A_1 is the adjacency matrix of X.

Case 2 **The representation ρ.**
Here we find

$$A_\rho = A(1) - A(\sigma) = \begin{pmatrix} 0 & 1 & -1 & -1 \\ 1 & 0 & 1 & 1 \\ -1 & 1 & 0 & -1 \\ -1 & 1 & -1 & 0 \end{pmatrix}.$$

Now we proceed to check our formulas for this case. We know by Corollary 18.11 that

$$\zeta_Y(u) = L(u, 1, Y/Y) = L(u, 1, Y/X)L(u, \rho, Y/X) = \zeta_X(u)L(u, \rho, Y/X). \tag{18.6}$$

The Ihara determinant formula, Theorem 2.5, implies that

$$\zeta_X(u)^{-1} = \left(1 - u^2\right)^2 (1 - u)(1 - 2u)\left(1 + u + 2u^2\right)^3$$

and

$$\zeta_Y(u)^{-1} = (1 - u^2)^2 (1 + u)(1 + 2u)(1 - u + 2u^2)^3 \zeta_X(u)^{-1}.$$

Then Theorem 18.15 implies (since $r = 3$) that

$$\begin{aligned} L(u, \rho, Y/X)^{-1} &= (1 - u^2)^2 \det\left(I_4 - A'_\rho u + 2u^2 I_4\right) \\ &= (1 - u^2)^2 (1 + u)(1 + 2u)(1 - u + 2u^2)^3. \end{aligned}$$

Note that $L(u, \rho, Y/X) = \zeta_X(-u)$, although this is not obvious from the determinant formula, where $-A \neq A_\rho$.

Note Theorem 18.15 implies that equation (18.6) is a factorization of an 8×8 determinant as a product of 4×4 determinants:

$$\det(I_8 - A_Y u + 2I_8 u^2) = \det(I_4 - A_X u + 2I_4 u^2)\, \det\big(I_4 - A'_\rho u + 2I_4 u^2\big).$$

Exercise 18.3 Compute the spectra of the adjacency matrices of the cube and the tetrahedron. Are these graphs Ramanujan?

It is perhaps worthwhile to state the case of 2-coverings separately. We will also give an example of this proposition below for the cube over the tetrahedron.

Proposition 18.17 *If Y/X is a 2-covering then the adjacency matrix A_Y has block decomposition with two blocks, one block being A_X (the adjacency matrix of X) and the other block being A_-. The matrix A_- is defined by having entries corresponding to two vertices a, b of X given by:*

$$(A_-)_{a,b} = \begin{cases} +1, & \text{a and b joined by edge e in X which lifts to edge of Y} \\ & \text{that does not change sheets;} \\ -1 & \text{a and b joined by edge e in X which lifts to edge of Y} \\ & \text{that changes sheets;} \\ 0 & \text{a and b not joined by edge e in X.} \end{cases}$$

Note the following conjecture made in Hoory *et al.* [55].

Conjecture 18.18 *Every d-regular graph X has a 2-covering Y such that if A_Y is the adjacency matrix of Y then*

$$\text{Spectrum } A_Y - \text{Spectrum } A_X \subset \left[-2\sqrt{d-1},\ 2\sqrt{d-1}\right].$$

This conjecture would allow one to construct families of Ramanujan graphs of arbitrary degree with number of vertices going to infinity by taking repeated 2-covers. By Proposition 18.17, this is a conjecture about the spectrum of the matrix A_-.

Exercise 18.4 (a) For Example 18.16 all edges of K_4 not in the chosen spanning tree were lifted to start on one sheet and end on the other sheet of Y. What happens if you consider the cover Y' in Exercise 2.7 instead?

(b) Check Conjecture 18.18 for the spectra of 2-covers of K_4.

Exercise 18.5 Experiment with Conjecture 18.18 concerning the spectra of 2-covers to see whether any k-regular graph does have a 2-cover such that the spectrum of A_- lies in the interval $[-2\sqrt{k-1}, 2\sqrt{k-1}]$. For example, you could look at all 2-coverings of any of our favorite graphs.

There are commands in Mathematica to do most of this.

Example 18.19: The cube over the dumbbell The covering that we consider in this example is Y/X in Figure 13.8. The covering group $G(Y/X)$ is the integers mod 4, denoted $\mathbb{Z}_4 = \{0, 1, 2, 3 \pmod 4\}$. We label the sheets as follows:

$$x_1' = (x, 0(\text{mod } 4)), \quad x_2' = (x, 1(\text{mod } 4)),$$
$$x_1'' = (x, 2(\text{mod } 4)), \quad x_2'' = (x, 3(\text{mod } 4)).$$

The irreducible representations are all one-dimensional and may be written $\chi_\nu(j) = \exp(2\pi i \nu j/4) = i^{\nu j}$ for $j, \nu \in \mathbb{Z}_4$. Note that although X has loops, Y does not. It follows that

$$A(0) = \begin{pmatrix} 0 & 1 \\ 1 & 0 \end{pmatrix}, \qquad A(1) = A(3) = I_2, \qquad A(2) = 0.$$

Thus

$$A_{\chi_0} = \begin{pmatrix} 2 & 1 \\ 1 & 2 \end{pmatrix}, \qquad A_{\chi_1} = \begin{pmatrix} 0 & 1 \\ 1 & 0 \end{pmatrix} = A_{\chi_3}, \qquad A_{\chi_2} = \begin{pmatrix} -2 & 1 \\ 1 & -2 \end{pmatrix}.$$

The corresponding L-functions are

$$\begin{aligned} L(u, \chi_0, Y/X)^{-1} &= (1 - u^2) \det \begin{pmatrix} 1 - 2u + 2u^2 & -u \\ -u & 1 - 2u + 2u^2 \end{pmatrix} \\ &= (1 - u^2)(1 - u)(1 - 2u)(1 - u + 2u^2); \end{aligned}$$

$$\begin{aligned} L(u, \chi_1, Y/X)^{-1} = L(u, \chi_3, Y/X) &= (1 - u^2) \det \begin{pmatrix} 1 + 2u^2 & -u \\ -u & 1 + 2u^2 \end{pmatrix} \\ &= (1 - u^2)(1 + u + 2u^2)(1 - u + 2u^2); \end{aligned}$$

$$\begin{aligned} L(u, \chi_2, Y/X)^{-1} &= (1 - u^2) \det \begin{pmatrix} 1 + 2u + 2u^2 & -u \\ -u & 1 + 2u + 2u^2 \end{pmatrix} \\ &= (1 - u^2)(1 + u)(1 + 2u)(1 + u + 2u^2). \end{aligned}$$

One sees again that as in Corollary 18.11,

$$\zeta_Y(u)^{-1} = L(u, \chi_0, Y/X)L(u, \chi_1, Y/X)L(u, \chi_2, Y/X)L(u, \chi_3, Y/X).$$

Note Again you can view the preceding equality as a factorization of the determinant of an 8×8 matrix into a product of four determinants of 2×2 matrices.

Example 18.20: An S_3 cover Now consider the example in Figure 14.4. Here we will view the group S_3 as the dihedral group D_3. Thus it consists of motions of a regular triangle and is generated by a flip F and a rotation R:

$$A(I) = \begin{pmatrix} 0 & 1 & 1 & 1 \\ 1 & 0 & 0 & 0 \\ 1 & 0 & 0 & 0 \\ 1 & 0 & 0 & 0 \end{pmatrix}, \qquad A(FR^2) = \begin{pmatrix} 0 & 0 & 0 & 0 \\ 0 & 0 & 0 & 0 \\ 0 & 0 & 0 & 1 \\ 0 & 0 & 1 & 0 \end{pmatrix},$$

$$A(FR) = \begin{pmatrix} 0 & 0 & 0 & 0 \\ 0 & 0 & 0 & 1 \\ 0 & 0 & 0 & 0 \\ 0 & 1 & 0 & 0 \end{pmatrix}, \qquad A(R^2) = 0, \qquad A(R) = 0, \qquad A(F) = 0.$$

Next we need to know the representations of S_3. See Example 18.20. The non-trivial one-dimensional representation of S_3 has the values $\chi_1(FR) = -1$ and $\chi_1(FR^2) = -1$. The two-dimensional representation ρ has the values

$$\rho(FR) = \begin{pmatrix} 0 & \omega^2 \\ \omega & 0 \end{pmatrix} \quad \text{and} \quad \rho\left(FR^2\right) = \begin{pmatrix} 0 & \omega \\ \omega^2 & 0 \end{pmatrix}, \qquad \omega = e^{2\pi i/3}.$$

Now we can compute the matrices in our L-functions:

$$A_{\chi_0} = A, \qquad A_{\chi_1} = \begin{pmatrix} 0 & 1 & 1 & 1 \\ 1 & 0 & 0 & -1 \\ 1 & 0 & 0 & -1 \\ 1 & -1 & -1 & 0 \end{pmatrix},$$

$$A_\rho = A_1(I) \otimes \rho(I) + A_1(FR) \otimes \rho(FR) + A_1(FR^2) \otimes \rho(FR^2)$$

$$= \begin{pmatrix} 0 & 1 & 1 & 1 & 0 & 0 & 0 & 0 \\ 1 & 0 & 0 & 0 & 0 & 0 & 0 & \omega^2 \\ 1 & 0 & 0 & 0 & 0 & 0 & 0 & \omega \\ 1 & 0 & 0 & 0 & 0 & \omega^2 & \omega & 0 \\ 0 & 0 & 0 & 0 & 0 & 1 & 1 & 1 \\ 0 & 0 & 0 & \omega & 1 & 0 & 0 & 0 \\ 0 & 0 & 0 & \omega^2 & 1 & 0 & 0 & 0 \\ 0 & \omega & \omega^2 & 0 & 1 & 0 & 0 & 0 \end{pmatrix}.$$

It follows that

$$L(u,\chi_0,Y_6/X)^{-1} = (1-u^2)\det\begin{pmatrix} 1+2u^2 & -u & -u & -u \\ -u & 1+u^2 & 0 & -u \\ -u & 0 & 1+u^2 & -u \\ -u & -u & -u & 1+2u^2 \end{pmatrix}$$

$$= (1-u^2)(1-u)(1+u^2)(1+u+2u^2)(1-u^2-2u^3),$$

$$L(u,\chi_1,Y_6/X)^{-1} = (1-u^2)\det\begin{pmatrix} 1+2u^2 & -u & -u & -u \\ -u & 1+u^2 & 0 & u \\ -u & 0 & 1+u^2 & u \\ -u & u & u & 1+2u^2 \end{pmatrix}$$

$$= (1-u^2)(1+u)(1+u^2)(1-u+2u^2)(1-u^2+2u^3),$$

$$L(u,\rho,Y_6/X)^{-1} = (1-u^2)^2\det(I_8 - A_\rho u + u^2 Q_\rho).$$

$$= (1-u^2)^2(1+u+2u^2+u^3+2u^4)(1+u+u^3+2u^4)$$
$$\times\,(1-u+2u^2-u^3+2u^4)(1-u-u^3+2u^4).$$

Putting all our results together and using Theorem 18.15, Exercise 18.1, and Proposition 18.10, we have

$$\zeta_X(u)^{-1} = L(u,\chi_0,Y_6/X)^{-1}$$
$$= (1-u^2)(1-u)(1+u^2)(1+u+2u^2)(1-u^2-2u^3),$$
$$\zeta_{Y_2}(u)^{-1}\zeta_X(u) = L(u,\chi_1,Y_2/X)^{-1} = L(u,\chi_1,Y_6/X)^{-1}$$
$$= (1-u^2)(1+u)(1+u^2)(1-u+2u^2)(1-u^2+2u^3),$$
$$\zeta_{Y_3}(u)^{-1}\zeta_X(u) = L(u,\rho,Y_6/X)^{-1}$$
$$= (1-u^2)^2(1+u+2u^2+u^3+2u^4)(1+u+u^3+2u^4)$$
$$\times\,(1-u+2u^2-u^3+2u^4)(1-u-u^3+2u^4),$$

and

$$\zeta_{Y_6}(u) = L(u,\chi_0,Y_6/X)L(u,\chi_1,Y_6/X)L(u,\rho,Y_6/X)^2$$
$$= \zeta_X(u)\frac{\zeta_{Y_2}(u)}{\zeta_X(u)}\left[\frac{\zeta_{Y_3}(u)}{\zeta_X(u)}\right]^2.$$

As a consequence, we find that

$$\zeta_X(u)^2\zeta_{Y_6}(u) = \zeta_{Y_2}(u)\zeta_{Y_3}(u)^2.$$

This is analogous to an example of zeta functions of pure cubic extensions of number fields that goes back to Dedekind. See Stark [118].

Note The last equality can be viewed as two different factorizations of determinants involving polynomials in u.

Example 18.21: The Klein 4-group cover Y/X from Figure 13.9 Here we can identify the Galois group $G = G(Y/X)$ with $\mathbb{Z}_2^2$. The identification is given by $x_1' = (x, (1,0))$, $x_1'' = (x, (1,1))$, $x_2' = (x, (0,0))$, $x_2'' = (x, (0,1))$.

The representations of G are given by $\chi_{r,s}(u, v) = (-1)^{ru+sv}$ for $r, s, u, v \in \mathbb{Z}_2$. We find that

$$L(u, \chi_{0,0}, Y/X)^{-1} = (1-u^2)\det\begin{pmatrix} 1+2u^2 & -3u \\ -3u & 1+2u^2 \end{pmatrix}$$
$$= Z_X(u)^{-1} = (1-u^2)(1-u)(1+u)(1-2u)(1+2u).$$

Similarly

$$L(u, \chi_{0,1}, Y/X)^{-1} = (1-u^2)\det\begin{pmatrix} 1+2u^2 & -u \\ -u & 1+2u^2 \end{pmatrix} = L(u, \chi_{1,1}, Y/X)^{-1}$$
$$= Z_X(u)^{-1} = (1-u^2)(1-u+2u^2)(1+u+2u^2)$$

and

$$L(u, \chi_{1,0}, Y/X)^{-1} = (1-u^2)\det\begin{pmatrix} 1+2u^2 & u \\ u & 1+2u^2 \end{pmatrix}$$
$$= (1-u^2)(1-u+2u^2)(1+u+2u^2).$$

Thus all three L-functions with non-trivial representations are equal. This happens here because all three intermediate quadratic covers of X are isomorphic as abstract graphs and so they have equal zeta functions. Each intermediate zeta function is of the form $\zeta_{\widetilde{X}}(u) = \zeta_X(u)L(u, \chi, Y/X)$, where χ runs through the three non-trivial representations of G as $\widetilde{X}$ runs through the three intermediate quadratic covers of X. For $\zeta_Y(u)$ we have

$$\zeta_Y(u)^{-1} = \prod_{\chi \in \widehat{G}} L(u, \chi, Y/X) = (1-u^2)^4(1-u)(1+u)(1-2u)$$
$$\times (1+2u)(1-u+2u^2)^3(1+u+2u^2)^3.$$

We also have

$$\zeta_X^2(u)\zeta_Y(u) = \zeta_{\widetilde{X}}(u)^3$$

which holds for all three intermediate quadratic covers $\widetilde{X}$ of X.

Example 18.22: The cyclic 6-fold cover Y/X from Figure 13.10 The covering group $G = G(Y/X) \cong \mathbb{Z}_6 = \{1, 2, 3, 4, 5, 6 \pmod 6\}$, with identity element 6(mod 6). Let $\omega = e^{2\pi i/6}$. The representations are $\chi_a(b) = \omega^{ab}$ for $a, b \in \mathbb{Z}_6$. Here the matrices $A(\tau)$ are 1×1. We obtain

$$A(6) = A(3) = 0, \qquad A(1) = A(2) = A(4) = A(5) = 1.$$

We find that

$$\begin{aligned} A_{\chi_0} &= 4 = A, \quad \text{the adjacency matrix of } X, \\ A_{\chi_j} &= 0 \quad \text{for } j = 1, 3, 5, \\ A_{\chi_j} &= -2 \quad \text{for } j = 2, 4. \end{aligned}$$

Then

$$\begin{aligned} L(u, \chi_0, Y/X)^{-1} &= \zeta_X(u)^{-1} = (1 - u^2)(1 - u)(1 - 3u), \\ L(u, \chi_j, Y/X)^{-1} &= (1 - u^2)(1 + 3u^2), \qquad \text{for } j = 1, 3, 5, \\ L(u, \chi_j, Y/X)^{-1} &= Z_X(u)^{-1} = (1 - u^2)(1 + 2u + 3u^2), \qquad \text{for } j = 2, 4. \end{aligned}$$

Set

$$m = \begin{pmatrix} 1+3u^2 & -u & -u & 0 & -u & -u \\ -u & 1+3u^2 & -u & -u & 0 & -u \\ -u & -u & 1+3u^2 & -u & -u & 0 \\ 0 & -u & -u & 1+3u^2 & -u & -u \\ -u & 0 & -u & -u & 1+3u^2 & -u \\ -u & -u & 0 & u & -u & 1+3u^2 \end{pmatrix}.$$

By Ihara's formula,

$$\begin{aligned} \zeta_Y(u)^{-1} &= \left(1 - u^2\right)^6 \det m \\ &= \left(1 - u^2\right)^6 \left(3u - 1\right)\left(u - 1\right)\left(3u^2 + 2u + 1\right)^2 \left(1 + 3u^2\right)^3, \end{aligned}$$

which agrees with the product

$$\zeta_Y(u)^{-1} = \prod_{\chi \in \widehat{G}} L(u, \chi, Y/X).$$

Exercise 18.6 Check whether the Ihara zetas of the preceding graphs and covering graphs satisfy the Riemann hypothesis.

19

Edge Artin L-functions

19.1 Definition and properties of edge Artin L-functions

Suppose that Y/X is a normal graph covering and recall Definitions 11.2, 11.3, and 16.3, of the edge matrix, edge zeta function, and Frobenius automorphism, respectively. We will use these definitions to define the edge Artin L-function, once again imitating the number theory Artin L-functions in Figure 16.1.

Definition 19.1 Given a path C in X which is written as a product of oriented edges $C = a_1 a_2 \cdots a_s$, the **edge norm** of C is

$$N_{\mathrm{E}}(C) = w_{a_1 a_2} w_{a_2 a_3} \cdots w_{a_{s-1} a_s} w_{a_s a_1}.$$

The **edge Artin L-function** associated to a representation ρ of the Galois group $G(Y/X)$ and the edge matrix W is

$$L(W, \rho) = L_{\mathrm{E}}(W, \rho, Y/X) = \prod_{[C]} \det\left(I - \rho\left(\frac{Y/X}{D}\right) N_{\mathrm{E}}(C)\right)^{-1},$$

where the product is over the primes $[C]$ in X and $[D]$ is arbitrarily chosen from the primes in Y over C. Here W is the edge matrix of Definition 11.2, with the variables $|w_{ef}|$ assumed sufficiently small, and $\left(\frac{Y/X}{D}\right)$ is the Frobenius automorphism of Definition 16.3.

Exercise 19.1 Show that the determinant in the definition of the edge zeta function does not depend on the choice of D over C in Definition 19.1.

Hint: The various Frobenius automorphisms are conjugate to each other.

For the factorization of edge zeta functions, we need a specialization of the W matrices.

Definition 19.2 Suppose that $\widetilde{X}$ is an unramified covering of X and that $\widetilde{W}$ and W are the corresponding edge matrices. Suppose that $\widetilde{e}$ and $\widetilde{f}$ are two edges of $\widetilde{X}$ with projections e and f in X using the covering map $\pi : \widetilde{X} \longrightarrow X$. If $\widetilde{e}$ feeds into $\widetilde{f}$ and $\widetilde{e} \neq \widetilde{f}^{-1}$ then e feeds into f and $e \neq f^{-1}$. Thus we can set the variable $\widetilde{w}_{\widetilde{e}\widetilde{f}} = w_{ef}$. When we do this for all the variables of $\widetilde{W}$, we call the X**-specialized edge matrix** $\widetilde{W}_{\text{spec}}$.

Theorem 19.3 (Main properties of edge Artin L-functions) *Assume that $\widetilde{X}$ is a normal (unramified) cover of X.*

(1) The edge L-function evaluated at the trivial representation is the usual edge zeta function:

$$L_{\text{E}}(W, 1, \widetilde{X}/X) = \zeta_{\text{E}}(W, X).$$

(2) The edge zeta function of $\widetilde{X}$, with the X-specialized edge matrix from Definition 19.2, factors as a product of edge L-functions:

$$\zeta_{\text{E}}(\widetilde{W}_{\text{spec}}, \widetilde{X}) = \prod_{\rho \in \widehat{G}} L_{\text{E}}(W, \rho)^{d_\rho}.$$

Here the product is over all inequivalent irreducible unitary representations of the Galois group $\text{Gal}(\widetilde{X}/X)$.

(3) Let $m = |E|$ be the number of unoriented edges of X. If the representation ρ of G has degree d, define the $2dm \times 2dm$ ***Artin edge matrix*** *W_ρ to have block form*

$$W_\rho = \big(w_{ef}\,\rho(\sigma(e))\big),$$

where $\sigma(e)$ denotes the normalized Frobenius element of Definition 16.1 corresponding to edge e. Then

$$L_{\text{E}}(W, \rho, Y/X) = \det(I - W_\rho)^{-1}.$$

Part (1) follows from the definitions. Part (2) is proved using Theorem 18.1 and parts (2) and (4) of the next theorem, as in the proof of Corollary 18.11 above. We will prove part (3) below in Section 19.2.

Theorem 19.4 (More properties of edge Artin L-functions) *Assume that Y/X is an (unramified) normal cover with Galois group G.*

(1) If you specialize the non-zero w_{ij} to be u then the Artin edge L-function $L_{\text{E}}(W, \rho)$ specializes to the Artin–Ihara L-function $L(u, \rho)$.

(2) $L_{\text{E}}(W, \rho_1 \oplus \rho_2) = L_{\text{E}}(W, \rho_1) L_{\text{E}}(W, \rho_2)$.

(3) *If* $\widetilde{X}$ *is intermediate to* Y/X, *then* $G = \mathrm{Gal}(Y/X)$ *and* $H = \mathrm{Gal}(Y/\widetilde{X})$. *Assume that* $\widetilde{X}/X$ *is normal. Let* ρ *be a representation of* $G/H \cong \mathrm{Gal}(\widetilde{X}/X)$. *We can view* ρ *as a representation of* G *(the* lift *of* ρ*). Then*

$$L_{\mathrm{E}}(W, \rho, Y/X) = L_{\mathrm{E}}(W, \rho, \widetilde{X}/X).$$

(4) *(Induction property) Suppose that* H *is any subgroup of* $G = \mathrm{Gal}(Y/X)$. *Let* $\widetilde{X}$ *be the cover intermediate to* Y/X *and corresponding to* H *by Theorem 14.3. Here we do not assume that* H *is a normal subgroup of* G. *Let* ρ *be a representation of* H *and let* $\rho^{\#}$ *denote the* representation of G induced by ρ. *Then, using Definition 19.2 of* $\widetilde{W}_{\mathrm{spec}}$,

$$L_{\mathrm{E}}(\widetilde{W}_{\mathrm{spec}}, \rho, Y/\widetilde{X}) = L_{\mathrm{E}}(W, \rho^{\#}, Y/X).$$

Proof Part (1) follows from the definitions. Part (2) is easily proved by rewriting the logarithm of the L-function as a sum involving traces of the representations, since $\mathrm{Tr}(\rho_1 \oplus \rho_2) = \mathrm{Tr}\ \rho_1 + \mathrm{Tr}\ \rho_2$. Part (3) follows from the definitions. Part (4) will be proved later, in Section 19.3. □

Example 19.5: Edge L-function of a cube covering a dumbbell The edge L-functions for the representations of the Galois group of Y/X, which is $\mathbb{Z}_4$, require the matrix W which has entry w_{ij} when edge e_i feeds into edge e_j. For the labeling of the edges of the dumbbell, see Figure 19.1. We find that the matrix W is given as in Example 11.8 by

$$W = \begin{pmatrix} w_{11} & w_{12} & 0 & 0 & 0 & 0 \\ 0 & 0 & w_{23} & 0 & 0 & w_{26} \\ 0 & 0 & w_{33} & 0 & w_{35} & 0 \\ 0 & w_{42} & 0 & w_{44} & 0 & 0 \\ w_{51} & 0 & 0 & w_{54} & 0 & 0 \\ 0 & 0 & 0 & 0 & w_{65} & w_{66} \end{pmatrix}.$$

Next we need to compute $\sigma(e_i)$ for each edge e_i, where $\sigma(C)$ denotes the normalized Frobenius automorphism of Definition 16.1. We will write the Galois group $G(Y/X)$ as $\{\sigma_1, \sigma_2, \sigma_3, \sigma_4\}$, where $(x, \sigma_j) = x^{(j)}$ for $x \in X$. The identification of $G(Y/X)$ with $\mathbb{Z}_4$ sends σ_j to $(j - 1 (\mathrm{mod}\, 4))$. Then we can compute the Galois group elements associated with the edges, $\sigma(e_1) = \sigma_2$, $\sigma(e_2) = \sigma_1$, $\sigma(e_3) = \sigma_2$. The representations of our group are one-dimensional and are given by $\chi_a(\sigma_b) = i^{a(b-1)}$ for $a, b \in \mathbb{Z}_4$.

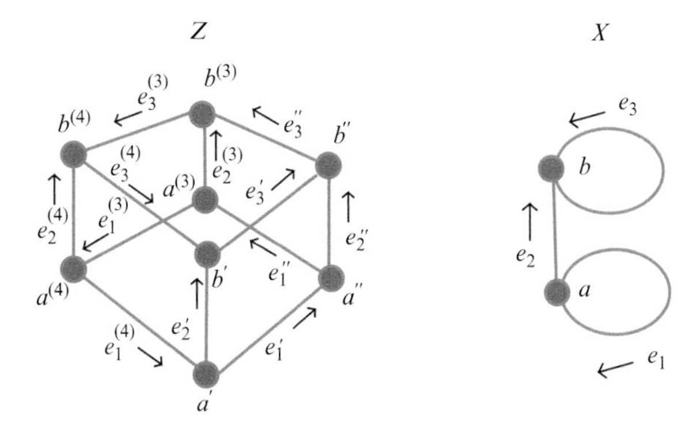

Figure 19.1 Edge labelings for the cube as a $\mathbb{Z}_4$ covering of the dumbbell.

So we obtain

$$L_{\mathrm{E}}(W, \chi_0, Y/X)^{-1} = \zeta_{\mathrm{E}}(W, X)^{-1}$$

$$= \det \begin{pmatrix} w_{11}-1 & w_{12} & 0 & 0 & 0 & 0 \\ 0 & -1 & w_{23} & 0 & 0 & w_{26} \\ 0 & 0 & w_{33}-1 & 0 & w_{35} & 0 \\ 0 & w_{42} & 0 & w_{44}-1 & 0 & 0 \\ w_{51} & 0 & 0 & w_{54} & -1 & 0 \\ 0 & 0 & 0 & 0 & w_{65} & w_{66}-1 \end{pmatrix},$$

$$L_{\mathrm{E}}(W, \chi_1, Y/X)^{-1} = \det(I - W_{\chi_1})$$

$$= \det \begin{pmatrix} iw_{11}-1 & iw_{12} & 0 & 0 & 0 & 0 \\ 0 & -1 & w_{23} & 0 & 0 & w_{26} \\ 0 & 0 & iw_{33}-1 & 0 & iw_{35} & 0 \\ 0 & -iw_{42} & 0 & -iw_{44}-1 & 0 & 0 \\ w_{51} & 0 & 0 & w_{54} & -1 & 0 \\ 0 & 0 & 0 & 0 & -iw_{65} & -iw_{66}-1 \end{pmatrix},$$

$$L_{\mathrm{E}}(W, \chi_2, Y/X)^{-1} = \det(I - W_{\chi_2})$$

$$= \det \begin{pmatrix} -w_{11}-1 & -w_{12} & 0 & 0 & 0 & 0 \\ 0 & -1 & w_{23} & 0 & 0 & w_{26} \\ 0 & 0 & -w_{33}-1 & 0 & -w_{35} & 0 \\ 0 & -w_{42} & 0 & -w_{44}-1 & 0 & 0 \\ w_{51} & 0 & 0 & w_{54} & -1 & 0 \\ 0 & 0 & 0 & 0 & -w_{65} & -w_{66}-1 \end{pmatrix},$$

$$L_{\mathrm{E}}(W,\chi_3,Y/X)^{-1} = \det(I - W_{\chi_3})$$

$$= \det\begin{pmatrix} -iw_{11}-1 & -iw_{12} & 0 & 0 & 0 & 0 \\ 0 & -1 & w_{23} & 0 & 0 & w_{26} \\ 0 & 0 & -iw_{33}-1 & 0 & -iw_{35} & 0 \\ 0 & iw_{42} & 0 & iw_{44}-1 & 0 & 0 \\ w_{51} & 0 & 0 & w_{54} & -1 & 0 \\ 0 & 0 & 0 & 0 & iw_{65} & iw_{66}-1 \end{pmatrix}.$$

Note that, by part (2) of Theorem 19.3, the product of the preceding four 6×6 determinants has to be a 24×24 determinant:

$$\det(I - \widetilde{W}_{\mathrm{spec}}) = \prod_{i=0}^{3} \det(I - W_{\chi_i}),$$

where $\widetilde{W}_{\mathrm{spec}}$ is found by specializing the edge matrix of the cube as follows. For the edge variables of the dumbbell, we write

$$a = w_{11}, \quad b = w_{12}, \quad c = w_{23}, \quad d = w_{26}, \quad e = w_{33}, \quad f = w_{35},$$
$$g = w_{42}, \quad h = w_{44}, \quad j = w_{51}, \quad k = w_{54}, \quad m = w_{65}, \quad n = w_{66}.$$

Using the natural ordering of the edges of Y given by $e_1', e_1'', e_1^{(3)}, e_1^{(4)}, e_2', e_2'', e_2^{(3)}, e_2^{(4)}, e_3', e_3'', e_3^{(3)}, e_3^{(4)}$, and the inverses in the same order, the matrix $\widetilde{W}_{\mathrm{spec}}$ is the following:

$$\begin{pmatrix}
0&a&0&0&0&b&0&0&0&0&0&0&0&0&0&0&0&0&0&0&0&0&0&0\\
0&0&a&0&0&0&b&0&0&0&0&0&0&0&0&0&0&0&0&0&0&0&0&0\\
0&0&0&a&0&0&0&b&0&0&0&0&0&0&0&0&0&0&0&0&0&0&0&0\\
a&0&0&0&b&0&0&0&0&0&0&0&0&0&0&0&0&0&0&0&0&0&0&0\\
0&0&0&0&0&0&0&0&c&0&0&0&0&0&0&0&0&0&0&0&0&0&0&d\\
0&0&0&0&0&0&0&0&0&c&0&0&0&0&0&0&0&0&0&0&d&0&0&0\\
0&0&0&0&0&0&0&0&0&0&c&0&0&0&0&0&0&0&0&0&0&d&0&0\\
0&0&0&0&0&0&0&0&0&0&0&c&0&0&0&0&0&0&0&0&0&0&d&0\\
0&0&0&0&0&0&0&0&0&e&0&0&0&0&0&0&0&f&0&0&0&0&0&0\\
0&0&0&0&0&0&0&0&0&0&e&0&0&0&0&0&0&0&f&0&0&0&0&0\\
0&0&0&0&0&0&0&0&0&0&0&e&0&0&0&0&0&0&0&f&0&0&0&0\\
0&0&0&0&0&0&0&0&e&0&0&0&0&0&0&0&f&0&0&0&0&0&0&0\\
0&0&0&0&g&0&0&0&0&0&0&0&0&0&0&h&0&0&0&0&0&0&0&0\\
0&0&0&0&0&g&0&0&0&0&0&0&h&0&0&0&0&0&0&0&0&0&0&0\\
0&0&0&0&0&0&g&0&0&0&0&0&0&h&0&0&0&0&0&0&0&0&0&0\\
0&0&0&0&0&0&0&g&0&0&0&0&0&0&h&0&0&0&0&0&0&0&0&0\\
j&0&0&0&0&0&0&0&0&0&0&0&0&0&0&k&0&0&0&0&0&0&0&0\\
0&j&0&0&0&0&0&0&0&0&0&0&k&0&0&0&0&0&0&0&0&0&0&0\\
0&0&j&0&0&0&0&0&0&0&0&0&0&k&0&0&0&0&0&0&0&0&0&0\\
0&0&0&j&0&0&0&0&0&0&0&0&0&0&k&0&0&0&0&0&0&0&0&0\\
0&0&0&0&0&0&0&0&0&0&0&0&0&0&0&0&m&0&0&0&0&0&0&n\\
0&0&0&0&0&0&0&0&0&0&0&0&0&0&0&0&0&m&0&0&n&0&0&0\\
0&0&0&0&0&0&0&0&0&0&0&0&0&0&0&0&0&0&m&0&0&n&0&0\\
0&0&0&0&0&0&0&0&0&0&0&0&0&0&0&0&0&0&0&m&0&0&n&0
\end{pmatrix}.$$

19.2 Proofs of determinant formulas for edge Artin L-functions

One can prove part (3) of Theorem 19.3 using the same method as we gave for the analogous result for the edge Ihara zeta function.

Proof of part (3) of Theorem 19.3 We are trying to prove that

$$L_{\mathrm{E}}(W, \rho, Y/X) = \det(I - W_\rho)^{-1}.$$

For this, we will imitate the proof of Theorem 18.8, setting $\chi = \mathrm{Tr}\ \rho$. So we start by taking the logarithm of the L-function, obtaining

$$\begin{aligned} \log(L_{\mathrm{E}}(W, \rho, Y/X)) &= \sum_{[P]} \sum_{j=1}^{\infty} \frac{1}{j} \chi(\sigma(P)^j) N_{\mathrm{E}}(P^j) \\ &= \sum_{P} \sum_{j=1}^{\infty} \frac{1}{j\nu(P)} \chi(\sigma(P^j)) N_{\mathrm{E}}(P^j). \end{aligned}$$

Now the sum is over paths P rather than classes $[P]$.

The block $i_1 i_{n+1}$ entry of W_ρ^n is (as we have noted in many earlier proofs)

$$\begin{aligned} &\sum_{i_2,\ldots,i_n} w(i_1, i_2) \cdots w(i_n, i_{n+1}) \rho(\sigma(i_1)) \cdots \rho(\sigma(i_n)) \\ &= \sum_{\substack{C = i_1 \cdots i_n \\ \nu(C) = n}} w(i_1, i_2) \cdots w(i_n, i_{n+1}) \rho(\sigma(C)), \end{aligned}$$

where the sum is over all paths C on X of length n with leading edge i_1.

The only non-zero entries in the last sum are for those paths C whose initial edge is i_1 and whose terminal edge i_n feeds into i_{n+1} without backtracking such that $i_{n+1} \neq i_n^{-1}$. Thus, when taking the trace, we have $i_{n+1} = i_1$ and we are talking about closed, backtrackless, tailless paths of length n.

Therefore using Exercise 4.1 as usual, we see that

$$\begin{aligned} \log L_{\mathrm{E}}(W, \rho, Y/X) &= \sum_{C} \frac{\mathrm{Tr}\ \rho(C)}{\nu(C)} N_{\mathrm{E}}(C) \\ &= \sum_{m \geq 1} \frac{1}{m} \mathrm{Tr}\ W_\rho^m = \mathrm{Tr}\left(\log(I - W_\rho)^{-1}\right) \\ &= \log \det\left((I - W_\rho)^{-1}\right). \end{aligned}$$

This completes the proof of part (3) of Theorem 19.3. □

19.2.1 Bass proof of Ihara theorem for Artin L-functions

Next we give the Bass proof of the Ihara theorem, Theorem 18.15, for Artin L-functions. We must first generalize the S, T matrices in Proposition 11.7. For a representation ρ of the Galois group $G(Y/X)$, let d_ρ be its degree (i.e., the size of the matrices $\rho(g)$). When we write the tensor product $B \otimes C$ for the $p \times p$ matrix B and the $r \times r$ matrix C, we mean the $pr \times pr$ matrix with block decomposition

$$B \otimes C = \begin{pmatrix} b_{11}C & \cdots & b_{1p}C \\ \vdots & \ddots & \vdots \\ b_{p1}C & \cdots & b_{pp}C \end{pmatrix}.$$

Definition 19.6 With the definitions of the start, terminal, and J matrices S, T, J, as in Proposition 11.7, define the **Artinized start, terminal, and J matrices** by

$$S_\rho = S \otimes I_{d_\rho}, \qquad T_\rho = T \otimes I_{d_\rho}, \qquad J_\rho = J \otimes I_{d_\rho}.$$

Definition 19.7 We define the $2md_\rho \times 2md_\rho$ block diagonal R-**matrix** R_ρ to be

$$R_\rho = \begin{pmatrix} \rho(\sigma(e_1)) & \cdots & 0 \\ \vdots & \ddots & \vdots \\ 0 & \cdots & \rho(\sigma(e_{2m})) \end{pmatrix} = \begin{pmatrix} U & 0 \\ 0 & U^{-1} \end{pmatrix}. \tag{19.1}$$

Here $e_1, \ldots, e_{2m}$ is our list of oriented edges of X, ordered in our usual way via formula (2.1). The second equality comes from the property $\sigma(e)^{-1} = \sigma(e^{-1})$ of the normalized Frobenius automorphism.

Recall Definition 18.7 of the Artinized edge adjacency matrix $W_{1,\rho}$.

Finally, recall A_ρ, the Artinized adjacency matrix associated with ρ as given in Definition 18.13. Using all the preceding definitions, we have the following proposition relating all the matrices.

Proposition 19.8 (Formulas involving $\rho, Q, W, A, R, S, T, J$)

(1) $W_{1,\rho} = R_\rho(W_1 \otimes I_d)$.
(2) $A_\rho = S_\rho R_\rho {}^{\mathrm{t}}T_\rho$.
(3) $S_\rho J_\rho = T_\rho,\ T_\rho J_\rho = S_\rho,\ Q_\rho + I_{nd_\rho} = S_\rho {}^{\mathrm{t}}S_\rho = T_\rho {}^{\mathrm{t}}T_\rho$.
(4) $W_{1,\rho} + R_\rho J_\rho = R_\rho {}^{\mathrm{t}}T_\rho S_\rho$.
(5) $(R_\rho J_\rho)^2 = I_{2|E|d}$.

Proof Part (1) To see this, just multiply the matrices in block form, setting $W_1 = B$ with entries $b_{e,f}$:

$$(W_{1,\rho})_{e,f} = \rho(\sigma(e))\,(W_1)_{ef}$$

$$= \left(\begin{pmatrix} \rho(\sigma(e_1)) & \cdots & 0 \\ \vdots & \ddots & \vdots \\ 0 & \cdots & \rho(\sigma(e_{2m})) \end{pmatrix} \begin{pmatrix} b_{e_1,e_1} I_{d_\rho} & \cdots & b_{e_1,e_{2m}} I_{d_\rho} \\ \vdots & \ddots & \vdots \\ b_{e_{2m},e_1} I_{d_\rho} & \cdots & b_{e_{2m},e_{2m}} I_{d_\rho} \end{pmatrix} \right)_{e,f}.$$

Part (2) Set $d = d_\rho$. Then we have

$$S_\rho R_\rho {}^{\mathrm{t}}T_\rho = (S \otimes I_d) \begin{pmatrix} \rho(\sigma(e_1)) & \cdots & 0 \\ \vdots & \ddots & \vdots \\ 0 & \cdots & \rho(\sigma(e_{2m})) \end{pmatrix} {}^{\mathrm{t}}(T \otimes I_d)$$

$$= \begin{pmatrix} S_{11} I_d & \cdots & S_{1,2m} I_d \\ \vdots & \ddots & \vdots \\ S_{n1} I_d & \cdots & S_{n,2m} I_d \end{pmatrix} \begin{pmatrix} \rho(\sigma(e_1)) & \cdots & 0 \\ \vdots & \ddots & \vdots \\ 0 & \cdots & \rho(\sigma(e_{2m})) \end{pmatrix} \begin{pmatrix} t_{11} I_d & \cdots & t_{n1} I_d \\ \vdots & \ddots & \vdots \\ t_{1,2m} I_d & \cdots & t_{n,2m} I_d \end{pmatrix}.$$

Now look at the block corresponding to the vertices a, b of X, and obtain

$$\left(S_\rho R_\rho {}^{\mathrm{t}}T_\rho\right)_{a,b} = \sum_e s_{a,e}\rho(\sigma(e))t_{b,e} = \sum_{g \in G} \rho(g) \sum_{e,\sigma(e)=g} s_{a,e}t_{b,e}$$
$$= \sum_{g \in G} (A(g))_{a,b}\rho(g) = (A_\rho)_{a,b}.$$

Here the last equality uses the definition of A_ρ, Definition 18.13. The result in part (2) follows.

Part (3) The proof proceeds by the following computation:

$$\left(S_\rho J_\rho\right)_{v,e} = \left(\begin{pmatrix} s_{11} I_d & \cdots & s_{1,2m} I_d \\ \vdots & \ddots & \vdots \\ s_{n1} I_d & \cdots & s_{n,2m} I_d \end{pmatrix} \begin{pmatrix} 0 & I_m \otimes I_d \\ I_m \otimes I_d & 0 \end{pmatrix} \right)_{v,e} = (T)_{v,e},$$

$$\left(S_\rho {}^{\mathrm{t}}S_\rho\right)_{a,b} = \sum_e s_{a,e} I_d s_{b,e} I_d = (\#\text{ edges out of } a) \times \delta_{a,b} I_d = (Q + I)_{a,b} I_d.$$

Part (4) We have

$$\left(R_\rho\,{}^{\mathrm{t}}T_\rho S_\rho\right)_{e,f}$$

$$= \left(\begin{pmatrix} \rho(\sigma(e_1)) & \cdots & 0 \\ \vdots & \ddots & \vdots \\ 0 & \cdots & \rho(\sigma(e_{2m})) \end{pmatrix} \begin{pmatrix} t_{11}I_d & \cdots & t_{n1}I_d \\ \vdots & \ddots & \vdots \\ t_{1,2m}I_d & \cdots & t_{n,2m}I_d \end{pmatrix} \begin{pmatrix} s_{11}I_d & \cdots & s_{1,2m}I_d \\ \vdots & \ddots & \vdots \\ s_{n1}I_d & \cdots & s_{n,2m}I_d \end{pmatrix} \right)_{e,f}$$

$$= \sum_{\substack{v \\ \stackrel{e}{\to} v \stackrel{f}{\to}}} \rho(\sigma(e)) t_{ve} s_{vf} I_d = \rho(\sigma(e))\,(W_1)_{e,f} + \rho(\sigma(e)) J_{e,f}.$$

In the summation, directed edge e feeds into vertex v while directed edge f leads out of that same vertex. The second term after the final equals sign is for the case $f = e^{-1}$, when $(W_1)_{e,f} = 0$.

Part (5) To prove this, just note that:

$$\left(\begin{pmatrix} U & 0 \\ 0 & U^{-1} \end{pmatrix} \begin{pmatrix} 0 & I \\ I & 0 \end{pmatrix} \right)^2 = \begin{pmatrix} 0 & U \\ U^{-1} & 0 \end{pmatrix}^2 = \begin{pmatrix} I & 0 \\ 0 & I \end{pmatrix}.$$

□

Next, we will prove the main formulas in the Bass proof.

Proposition 19.9 (Main formulas in Bass proof of Ihara's theorem for Artin L-functions)

(1)
$$\begin{pmatrix} I_{nd} & 0 \\ R_\rho\,{}^tT_\rho & I_{2md} \end{pmatrix} \begin{pmatrix} I_{nd}\left(1-u^2\right) & S_\rho u \\ 0 & I_{2md} - uW_{1,\rho} \end{pmatrix}$$
$$= \begin{pmatrix} I_{nd} - A_\rho u + Q_\rho u^2 & S_\rho u \\ 0 & I_{2md} + R_\rho J_\rho u \end{pmatrix} \begin{pmatrix} I_{nd} & 0 \\ R_\rho\,{}^tT_\rho - {}^tS_\rho u & I_{2md} \end{pmatrix}.$$

(2)
$$I_{2md} + R_\rho J_\rho u = \begin{pmatrix} I_{md} & Uu \\ U^{-1}u & I_{md} \end{pmatrix}.$$

(3)
$$\begin{pmatrix} I_{md} & 0 \\ -U^{-1}u & I_{md} \end{pmatrix} \left(I_{2md} + R_\rho J_\rho u\right) = \begin{pmatrix} I_{md} & Uu \\ 0 & I_{md}\left(1-u^2\right) \end{pmatrix}.$$

Proof The proofs of parts (1)–(3) are **exercises** in the block multiplication of matrices using Proposition 19.8. □

Proof of Theorem 18.15 We are trying to show that the Artin-Ihara L-function has a three-term determinant formula:

$$L(u,\rho,Y/X)^{-1} = (1-u^2)^{(r-1)d}\det\big(I_{nd} - A_\rho u + Q_\rho u^2\big).$$

First recall Theorem 18.8:

$$L\,(u,\rho,Y/X)^{-1} = \det\big(I - uW_{1,\rho}\big)$$

Thus we need to show that

$$\det\big(I_{2md} - uW_{1,\rho}\big) = \big(1-u^2\big)^{(r-1)d}\det\big(I_{nd} - A_\rho u + Q_\rho u^2\big).$$

We see, upon taking determinants of the formula in part (1) of Proposition 19.9, that

$$\big(1-u^2\big)^{nd}\det\big(I_{2md} - uW_{1,\rho}\big)$$
$$= \det\big(I_{nd} - A_\rho u + Q_\rho u^2\big)\det\big(I_{2md} + R_\rho J_\rho u\big).$$

Then parts (2) and (3) of Proposition 19.9 imply that

$$\det(I_{2md} + R_\rho J_\rho u) = (1-u^2)^{md}.$$

Theorem 18.15 follows from the fact that $m - n = r - 1$. □

19.3 Proof of the induction property

The induction property of the edge Artin L-function, part (4) of Theorem 19.4 is the next thing for us to prove. To imitate the proof of the analogous number theory fact, one needs the following lemma.

Lemma 19.10 *Suppose that Y/X is normal with Galois group G and that H is the subgroup of G corresponding to an intermediate covering $\widetilde{X}$. Let $\chi = \mathrm{Tr}\,\rho$ be a character of a representation of H and $\chi^{\#} = \mathrm{Tr}\,\mathrm{Ind}_H^G\,\rho$ be the corresponding induced character of G. For any prime $[C]$ of X, we have*

$$\sum_{j=1}^{\infty}\frac{1}{j}\,\chi^{\#}\big(\sigma(C^j)\big)N_{\mathrm{E}}(C)^j = \sum_{[\widetilde{C}]|[C]}\sum_{j=1}^{\infty}\frac{1}{j}\,\chi\big(\widetilde{\sigma}(\widetilde{C})^j\big)N_{\mathrm{E}}(\widetilde{C})^j_{\mathrm{spec}}. \qquad (19.2)$$

Here $\sigma(C) \in G$ is the normalized Frobenius automorphism for C in X and $\widetilde{\sigma}(\widetilde{C}) \in H$ is the normalized Frobenius corresponding to $\widetilde{C}$ in $\widetilde{X}$. The X-specialized edge matrix in the norm on the right is from Definition 19.2.

Proof Let D_1 be the prime of Y above C starting on sheet 1. Then $\sigma(C) = [Y/X, D_1]$. Using the Frobenius formula for the induced character (Theorem 18.2), we have

$$\sum_{j=1}^{\infty} \frac{1}{j} \chi^{\#}\big(\sigma(C)^j\big) N_{\mathrm{E}}(C)^j$$

$$= \sum_{j=1}^{\infty} \sum_{\substack{g \in G \\ (g\sigma(C)g^{-1})^j \in H}} \frac{1}{j|H|} \chi\big((g\sigma(C)g^{-1})^j\big) N_{\mathrm{E}}(C)^j.$$

Each distinct prime $[D]$ of Y above C has the form $D = g \circ D_1$ and occurs for $f = f(D, Y/X)$ elements of G, where f is the residual degree of Definition 15.1. From Proposition 16.5 we see that

$$\sum_{j=1}^{\infty} \frac{1}{j|H|} \sum_{\substack{g \in G \\ (g\sigma(C)g^{-1})^j \in H}} \chi\big((g\sigma(C)g^{-1})^j\big) N(C)^j$$

$$= \sum_{[D] \big| [C]} \sum_{\substack{j \geq 1 \\ [Y/X, D]^j \in H}} \frac{f}{j|H|} \chi([Y/X, D]^j) N(C)^j.$$

Group the various D over C into those over a fixed $\widetilde{C}$ and then sum over the $\widetilde{C}$. For a fixed $\widetilde{C}$, all D dividing $\widetilde{C}$ have the same minimal power $j = f_1 = f(\widetilde{C}, \widetilde{X}/X)$ such that $[Y/X, D]^j \in H$. This power gives the Frobenius automorphism of D with respect to $Y/\widetilde{X}$, by Theorem 16.6.Thus the last double sum equals

$$\sum_{[\widetilde{C}] \big| [C]} \sum_{[D] \big| [\widetilde{C}]} \sum_{j \geq 1} \frac{f}{f_1 j |H|} \chi\big([Y/\widetilde{X}, D]^j\big) N(C)^{f_1 j}.$$

For all $[D] \big| [\widetilde{C}]$, the $[Y/\widetilde{X}, D]$ are conjugate to each other in H and there are g_2 such D, where $g_2 f_2 = |H|$. Here $f_2 = f(D, Y/\widetilde{X})$ and $g_2 = g(D, Y/\widetilde{X})$. If we pick one particular D above $\widetilde{C}$, we therefore get

$$\sum_{[D] \| [\widetilde{C}]} \sum_{j \geq 1} \frac{f}{f_1 j |H|} \chi\big([Y/\widetilde{X}, D]^j\big) N_{\mathrm{E}}(C)^{f_1 j}$$

$$= \sum_{j \geq 1} \frac{f g_2}{f_1 j |H|} \chi\big([Y/\widetilde{X}, D]^j\big) N(C)^{f_1 j}$$

$$= \sum_{j \geq 1} \frac{1}{j} \chi\big([Y/\widetilde{X}, D]^j\big) N(C)^{f_1 j}.$$

The proof is completed by putting the chain of equalities together, since

$$N(C)^{f_1} = N(\widetilde{C})_{\text{spec}}.$$

□

The following corollary will be needed for our discussion of graphs that are isospectral but not isomorphic.

Corollary 19.11 *Suppose that Y/X is normal with Galois group G and H is the subgroup of G corresponding to an intermediate cover $\widetilde{X}$. Let $\chi_1^\#$ be the character of the representation of G induced from the trivial representation* 1 *of H. Then the number of primes $[\widetilde{C}]$ of $\widetilde{X}$ above a prime $[C]$ of X with length $\nu(\widetilde{C}) = \nu(C)$ is $\chi_1^\#(\sigma(C))$, where $\sigma(C)$ denotes the normalized Frobenius automorphism from Definition 16.1. This means that $\chi_1^\#(\sigma(C))$ is the number of primes of $\widetilde{X}$ above $[C]$ with residual degree* 1.

Proof Set $\chi = \chi_1$ in Lemma 19.10 and set each non-zero edge variable w_{ij} equal to u. This makes $N_{\mathrm{E}}(C) = u^{\nu(C)}$ and $N_{\mathrm{E}}(\widetilde{C})_{\text{spec}} = u^{\nu(\widetilde{C})}$. Look at the $u^{\nu(C)}$ terms on each side of equation (19.2). The coefficient of $u^{\nu(C)}$ on the left-hand side comes from the $j = 1$ term and it is $\chi^\#(\sigma(C))$. The coefficient of $u^{\nu(C)}$ on the right-hand side is the number of $[\widetilde{C}]$ above $[C]$ with $\nu(\widetilde{C}) = \nu(C)$ since $\chi_1 = 1$. □

We can now give the proof we require.

Proof of the induction property of edge L-functions By the definition of the edge L-function for Y/X, we have

$$\log(L_{\mathrm{E}}(W, \rho^\#, Y/X)) = \sum_{[C]} \sum_{j=1}^{\infty} \frac{1}{j} \chi^\#(\sigma(C)^j) N(C)^j.$$

Apply Lemma 19.10 to see that the right-hand side is

$$\sum_{[\widetilde{C}]} \sum_{j=1}^{\infty} \frac{1}{j} \chi\big(\widetilde{\sigma}(\widetilde{C})^j\big) N_{\mathrm{E}}(\widetilde{C})^j_{\text{spec}},$$

where the sum is over all primes $\widetilde{C}$ of $\widetilde{X}$ and $\widetilde{\sigma}(\widetilde{C})$ is the corresponding normalized Frobenius automorphism in H. The proof is completed using the definition of the edge L-function for $Y/\widetilde{X}$. □

We could also give a purely combinatorial proof of the induction property – noting that the two determinants arising from part (3) of Theorem 19.3 are the

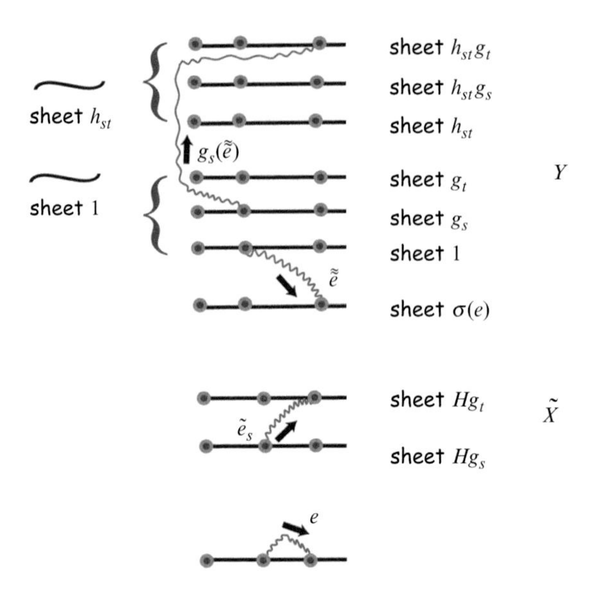

Figure 19.2 Proving the induced representation property of edge L-functions.

same size. Using the definition of induced representations, one can see that the two determinants are the same, as in the following exercise. See Stark and Terras [121] for some hints.

Exercise 19.2 Find a combinatorial proof of the induction property of the edge L-function,

$$L_{\mathrm{E}}(\widetilde{W}_{\mathrm{spec}}, \rho, Y/\widetilde{X}) = L_{\mathrm{E}}(W, \rho^{\#}, Y/X),$$

by looking at the formula $L_{\mathrm{E}}(W, \rho, Y/X) = \det(I - W_\rho)^{-1}$ and the analogous result for $L_{\mathrm{E}}(W, \rho^{\#}, Y/X)$, with $\rho^{\#} = \mathrm{Ind}_H^G \rho$.

Look at Figure 19.2. You need to split the $\widetilde{W}_\rho$ matrix into $2m \times 2m$ blocks indexed by the oriented edges e, f of X. The e block row comes from directed edges of $\widetilde{X}$ projecting to e. Edge e lifts to each sheet of $\widetilde{X}$. These sheets are labeled by cosets Hg_k. Label an edge of $\widetilde{X}$ as $\widetilde{e_s}$ if it projects to e and has its initial vertex on sheet Hg_s. Then we claim that

$$\left(\widetilde{W}_{\rho,\mathrm{spec}}\right)_{e,f} = \rho^{\#}(\sigma(e))\, w_{ef}.$$

To see this, suppose that edge $\widetilde{f_t}$ has its initial vertex on sheet Hg_t. If $\widetilde{e_s}$ feeds into $\widetilde{f_t}$, we see that e lifts to the edge of Y starting on sheet g_s and ending on sheet $h_{st}g_t$, with $h_{st} = \widetilde{\sigma}(\widetilde{e_s}) \in G(Y/\widetilde{X})$, where $\widetilde{\sigma}$ is the normalized Frobenius for $Y/\widetilde{X}$. In addition, we have

$$g_s\sigma(e) = h_{st}g_t$$

where $\sigma(e)$ is the normalized Frobenius for Y/X.

So we find that, extending ρ to be 0 outside H,

$$\left(\widetilde{W}_{\rho,\text{spec}}\right)_{e,f} = \left(\rho\left(g_s\sigma(e)g_t^{-1}\right)\right)w_{ef}.$$

Finish the proof with the formula for the matrix entries of an induced representation from Terras [133], p. 270.

20

Path Artin L-functions

20.1 Definition and properties of path Artin L-functions

There is one final kind of Artin L-function: the path L-function invented by Stark, which generalizes the path zeta function discussed earlier. Recall Definitions 12.1 and 12.2 of the path matrix Z, path norm, and path zeta function.

Definition 20.1 Assume Y/X normal with Galois group G. Given a representation ρ of G and path matrix Z with $|z_{ef}|$ sufficiently small, the **path Artin L-function** is defined by

$$L_{\mathrm{P}}(Z,\rho) = \prod_{\substack{[C]\ prime \\ in\ X}} \det\left(1 - \rho\left(\frac{Y/X}{D}\right) N_{\mathrm{P}}(C)\right)^{-1}.$$

Here $\left(\frac{Y/X}{D}\right)$ is from Definition 16.3, the path matrix Z is from Definition 12.1, the path norm $N_{\mathrm{P}}(C)$ is from Definition 12.2, and the product is over primes $[C]$ of X, with $[D]$ any prime of Y over $[C]$.

The path Artin L-function has analogous properties to the edge L-function. You just have to replace the subscript E with the subscript P in Theorems 19.3 and 19.4.

Proposition 20.2 (Some properties of the path Artin L-function)

(1) $L_{\mathrm{P}}(Z, 1, Y/X) = \zeta_{\mathrm{P}}(Z, X)$.

(2) $L_{\mathrm{P}}(Z, \rho_1 \oplus \rho_2, Y/X) = L_{\mathrm{P}}(Z, \rho_1, Y/X) L_{\mathrm{P}}(Z, \rho_2, Y/X)$.

(3) Let Y/X be normal with Galois group G, and let $\widetilde{X}$ be intermediate to Y/X such that $\widetilde{X}/X$ is normal, with $H = G(Y/\widetilde{X})$. Let ρ be a representation of $G/H \cong G(\widetilde{X}/X)$. View ρ as a representation of G (the lift of ρ). Then

$$L_{\mathrm{P}}(Z, \rho, Y/X) = L_{\mathrm{P}}(Z, \rho, \widetilde{X}/X).$$

Theorem 20.3 (Path Artin L-function as inverse of a polynomial) *The path L-function satisfies*

$$L_{\mathrm{P}}(Z, \rho, Y/X) = \det\left(I - Z_{\rho}\right)^{-1},$$

where the ***Artinized path matrix*** $Z_{\rho} = (z_{ef}\,\rho\,(\sigma(e)))$ *and* I *is the* $2dr \times 2dr$ *identity matrix, with* d *the degree of* ρ.

Proof The proof is like that of part (3) of Theorem 19.3 for the edge L-function. □

Just as with the path zeta functions the variables of the path L-function can be specialized to obtain the edge L-function. This specialization was given in formula (12.1).

Via this specialization, we find that

$$L_{\mathrm{P}}(Z(W), \rho) = L_E(W, \rho). \tag{20.1}$$

Example 20.4: Path Artin L-functions for a cyclic cover of two loops with an extra vertex on one loop Consider the base graph that has two loops with an extra vertex on one loop. Now, for our n-cyclic cover we lift edge a up one sheet and keep edge b in the same sheet. See Figure 20.1.

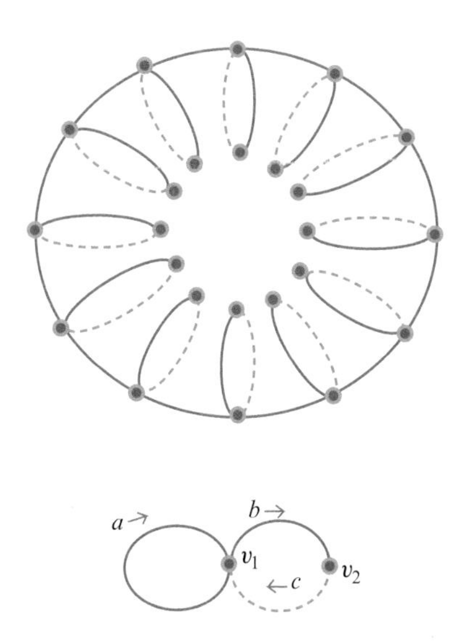

Figure 20.1 A **12-cyclic cover** of the base graph with two loops and an extra vertex on one loop. The spanning tree in the base graph is shown by a dashed line, as are the sheets of the cover above it.

The path matrix of the Artin L-function for an n-cyclic cover of two loops with an extra vertex on one loop as in Figure 20.1 is 4×4, and we can compute the L-functions for the n-cyclic cover by hand or use a computer. I used Scientific Workplace on my PC. So the L-function for the cyclic n-cover of the graph X consisting of two loops with an extra vertex on one loop in Figure 20.1, with $\rho = e^{2\pi i a/n}$ and $s = 2\cos(2\pi a/n)$, is

$$L(u, \chi_a)^{-1} = \det \begin{pmatrix} \rho u - 1 & \rho u & 0 & \rho u^2 \\ u^2 & u^2 - 1 & u^2 & 0 \\ 0 & \rho^{-1} u & \rho^{-1} u - 1 & \rho^{-1} u^2 \\ u & 0 & u & u^2 - 1 \end{pmatrix}$$

$$= \left(u^2 - 1\right)\left(-3u^4 + su^3 + su - 1\right).$$

When the character $\chi_a = 1$ we have $\rho = 1$, and we obtain the Ihara zeta function of the lower graph in Figure 20.1.

We can also compute the Artin L-functions of this cover using the 6×6 $W_{1,\rho}$ matrix. We decided to plot the eigenvalues of the matrices $W_{1,\rho}$. These eigenvalues are the reciprocals of the poles of the Ihara zeta of the cover. The result is shown in Figure 20.2, and it implies that the Riemann hypothesis is very false for this graph.

Exercise 20.1 Do the same experiment as in Example 20.4 but instead of keeping the lifts of edge b in the same sheet lift them down one sheet.

Exercise 20.2 Do the same experiment as in Example 20.4 but replace the base graph with $K_4 - e$, where e is an edge.

It is interesting to compare the spectra for the cyclic covers with those for other abelian covers of the same base graph as well as with those for random covers. See Figures 26.8 and 26.6.

20.2 Induction property

Next we want to discuss the induction property of path L-functions. For this, if $\widetilde{X}$ is a covering of X, we need to specialize the path matrix $\widetilde{Z}$ of $\widetilde{X}$ to the variables in the path matrix Z of X. This must be done in such a way that if $\widetilde{C}$ is a reduced cycle in its conjugacy class of the fundamental group of $\widetilde{X}$ then, under the specialization, $N_P(\widetilde{C})$ becomes $N_P(C)$, where C is the projected cycle of $\widetilde{C}$ in X.

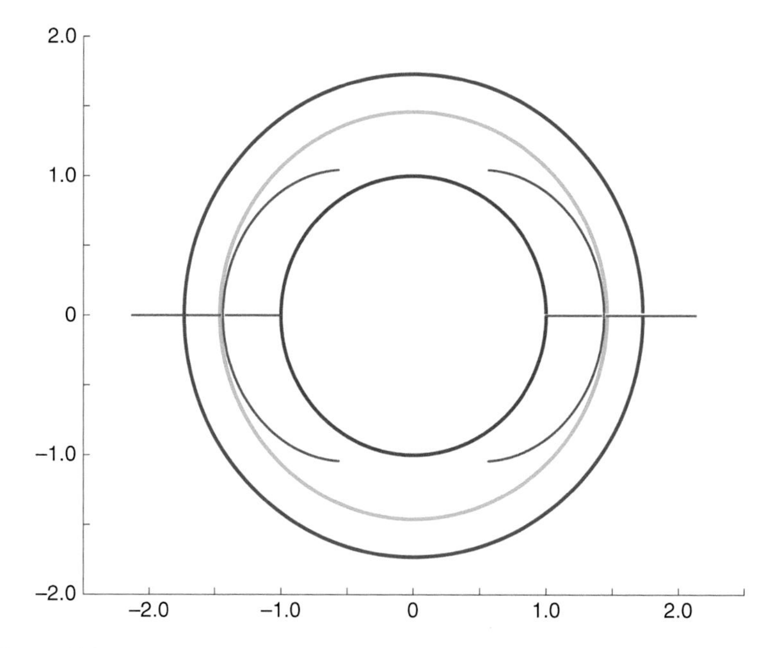

Figure 20.2 The purple points give the eigenvalues of the edge adjacency matrix W_1 for the 10 000 cyclic cover of the graph X consisting of two loops with an extra vertex on one loop analogous to the cover in Figure 20.1. These points are the reciprocals of the poles of the Ihara zeta function for the covering graph. The three circles are centered at the origin and have radii $\sqrt{p}$, $1/\sqrt{R}$, $\sqrt{q}$. Here $p = 1$, $1/R \cong 2.1304$, $q = 3$. The Riemann hypothesis is very false in this case.

Next we give the **specialization rule** for the induction property of path Artin L-functions. First we need a contraction rule. In X we contract the spanning tree T in the base graph X to a point. See Figure 20.3. This gives a graph $B(X)$ which is a bouquet of loops, pictured at lower right in the figure. The graphs X and $B(X)$ have the same fundamental group. The path and edge zeta functions of $B(X)$ are the same. In the cover $\widetilde{X}$ we also contract each sheet (the connected inverse images of T) to a point. This gives a graph $C(\widetilde{X})$, pictured at top right of Figure 20.3. The lifts to $\widetilde{X}$ of the r generating paths for the fundamental group of X give the edges of $C(\widetilde{X})$. What makes this interesting is that if $\widetilde{X}$ is an d-fold covering of X then $d - 1$ of the lifted edges from $B(X)$ must be used in the tree of $\widetilde{X}$. The remaining $dr - (d - 1) = d(r - 1) + 1$ non-tree edges of $C(\widetilde{X})$ give the generators of the fundamental group of $\widetilde{X}$. The specialization algorithm needs to take account of the tree edges.

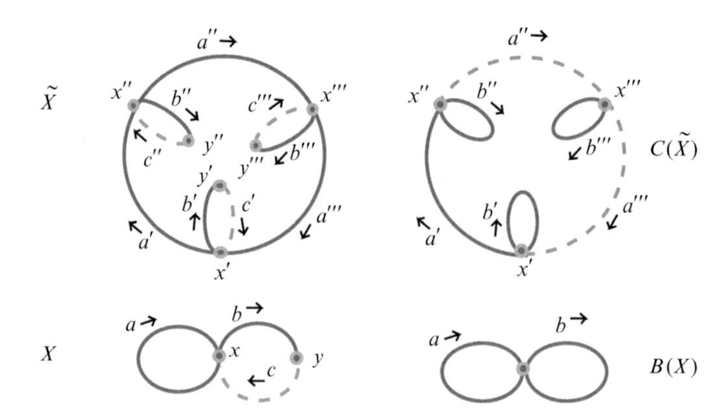

Figure 20.3 The contraction of sheets of a cover corresponding to contraction of a spanning tree. At upper left we have a $d = 3$-cyclic cover $\widetilde{X}$ of X. A spanning tree of X is shown at lower left by the dashed line. When we contract this spanning tree of X, we get the bouquet of loops $B(X)$ at lower right. At upper right the graph $C(\widetilde{X})$ is obtained by contracting the sheets of $\widetilde{X}$. In $C(\widetilde{X})$ the new spanning tree is shown by the dashed lines, and it has $d - 1 = 2$ edges.

First we specialize variables in the path matrix $\widetilde{Z}(\widetilde{X})$ to the edge variables on the contracted graph $C(\widetilde{X})$. This turns the path norm into the edge norm on $C(\widetilde{X})$. Then we specialize the edge variables of $C(\widetilde{X})$ to the edge variables of the contracted base graph $B(X)$ in our usual manner using the induction theorem for the edge Artin L-function. This turns the edge norm on $\widetilde{X}$ into the edge norm on $B(X)$, which is the same as the path norm on X. This is the desired specialization. Call it $\widetilde{Z}_{\text{spec}}$.

Example 20.5: Contracted covers The contracted versions of X and Y_3 from Figure 14.4 are shown in Figure 20.4.

We will use the following notation for the inverses of edges c and d: we write $C = c^{-1}$ and $D = d^{-1}$.

The tree $\widetilde{T}$ of Y_3 is completed with one of the lifts of the cC pair between the top two sheets of Y_3 and one of the lifts of the dD pair between the bottom two sheets. The remaining four undirected edges of the contracted Y_3 give rise to the fundamental group of Y_3 and the resulting 8×8 path matrix $\widetilde{Z}$. We give these edges directions projecting to either c or d rather than C or D, and labels I, II, III, IV, as shown. The inverse edges, projecting to C and D, are given labels $V, VI, VII, VIII$, as shown. The rows and columns of $\widetilde{Z}$ are then labeled by *I–VIII*.

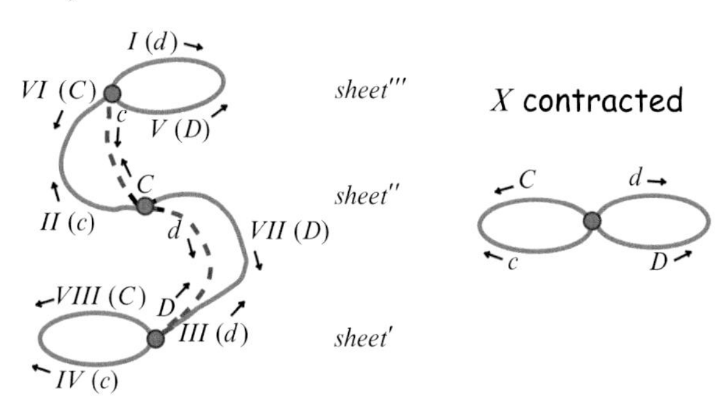

Figure 20.4 Contracted versions of X and Y_3 from Figure 14.4. The solid edges are the non-tree edges generating the fundamental group.

Following our specialization algorithm described above, the resulting specialized matrix $\widetilde{Z}_{\text{spec}}$ is then

$$\begin{pmatrix}
z_{dd} & z_{dc}z_{cc} & z_{dc}z_{cd}z_{dd} & z_{dc}z_{cd}z_{dc} & 0 & z_{dC} & z_{dc}z_{cD} & z_{dc}z_{cd}z_{dC} \\
z_{cd} & z_{cc}z_{cc} & z_{cc}z_{cd}z_{dd} & z_{cc}z_{cd}z_{dc} & z_{cD} & 0 & z_{cc}z_{cD} & z_{cc}z_{cd}z_{dC} \\
z_{dC}z_{Cd} & z_{dc} & z_{dd}z_{dd} & z_{dd}z_{dc} & z_{dC}z_{CD} & z_{dC}z_{CC} & 0 & z_{dd}z_{dC} \\
z_{cD}z_{DC}z_{Cd} & z_{cD}z_{Dc} & z_{cd} & z_{cc} & z_{cD}z_{DC}z_{CD} & z_{cD}z_{DC}z_{CC} & z_{cD}z_{DD} & 0 \\
0 & z_{Dc}z_{cc} & z_{Dc}z_{cd}z_{dd} & z_{Dc}z_{cd}z_{dc} & z_{DD} & z_{DC} & z_{Dc}z_{cD} & z_{Dc}z_{cd}z_{dC} \\
z_{CC}z_{Cd} & 0 & z_{Cd}z_{dd} & z_{Cd}z_{dc} & z_{CC}z_{CD} & z_{CC}z_{CC} & z_{CD} & z_{Cd}z_{dC} \\
z_{DD}z_{DC}z_{Cd} & z_{DD}z_{Dc} & 0 & z_{Dc} & z_{DD}z_{DC}z_{CD} & z_{DD}z_{DC}z_{CC} & z_{DD}z_{DD} & z_{DC} \\
z_{CD}z_{DC}z_{Cd} & z_{CD}z_{Dc} & z_{Cd} & 0 & z_{CD}z_{DC}z_{CD} & z_{CD}z_{DC}z_{CC} & z_{CD}z_{DD} & z_{CC}
\end{pmatrix}$$

For example, the IV, I entry $\widetilde{Z}_{\text{spec}\,IV,I}$ follows the directed edge IV (projecting to c) through two edges of $\widetilde{T}$ (projecting to D and C consecutively) to edge I (projecting to d), resulting in the specialized value $z_{cD}z_{DC}z_{Cd}$. This agrees with the fact that any path on Y_3 going through consecutive cut edges IV and I must project to a path on X going consecutively through c, D, C, d.

Exercise 20.3 Work out $\widetilde{Z}_{\text{spec}}$ for the example in Figure 20.3.

Theorem 20.6 (Induction property for path L-functions) *Suppose that Y/X is normal with Galois group G. If H is a subgroup of G corresponding to the intermediate covering $\widetilde{X}$, ρ is a representation of H, and $\rho^{\#}$ is the representation of G induced by ρ then, assuming that the variables of the path matrix $\widetilde{Z}$ for $Y/\widetilde{X}$ are specialized according to the specialization rule above, we have*

$$L_{\mathrm{P}}\big(\widetilde{Z}_{\text{spec}}, \rho, Y/\widetilde{X}\big) = L_{\mathrm{P}}(Z, \rho^{\#}, Y/X).$$

Proof Contract each copy of the tree T to a point, both in X and in $\widetilde{X}$. Then both sides of the equality in this theorem become edge L-functions attached to a graph with one vertex and r loops and its corresponding covering. Since the induction theorem was proved in Section 19.3 for edge L-functions, we are done. □

Remark From Theorem 20.3, the equality in Theorem 20.6 becomes

$$\det\left(I - \widetilde{Z}_{\mathrm{spec},\rho}\right) = \det(I - Z_{\rho^\#}).$$

Unlike the analogous equality for the edge L-functions, obtained from combining Theorems 19.3 and 19.4, here the determinants can have different sizes!

Corollary 20.7 (Factorization of the path zeta function) *Suppose that Y/X is normal with Galois group G. Then the path zeta function, once the variables are specialized as in Theorem 20.6, factors into a product of Artin L-functions:*

$$\zeta_{\mathrm{P}}\left(\widetilde{Z}_{\mathrm{spec}}, Y\right) = \prod_{\rho \in \widetilde{G}} L_{\mathrm{P}}(Z, \rho, Y/X)^{d_\rho}.$$

Proof The proof is the same as that for the analogous property for edge Artin L-functions. □

Example 20.8: Factorization of the path zeta function of a non-normal cubic cover Y_3 over X from the S_3 cover in Figure 14.4 Recall Example 18.20. Here we reconsider the example in light of the factorization theorem above for path zetas. We set $\omega = e^{2\pi i/3}$ and, using our labeling of edges from Figure 14.4, write

$$u_1 = z_{cc},\quad u_2 = z_{cd},\quad u_3 = z_{cD},\quad u_4 = z_{dc},\quad u_5 = z_{dd},\quad u_6 = z_{dC},$$

$$u_7 = z_{Cd},\quad u_8 = z_{CC},\quad u_9 = z_{CD},\quad u_{10} = z_{Dc},\quad u_{11} = z_{DC},\quad u_{12} = z_{DD}.$$

We find that, in an analogous manner to Example 18.20, the product of

$$\det\begin{pmatrix} u_1 - 1 & u_2 & 0 & u_3 \\ u_4 & u_5 - 1 & u_6 & 0 \\ 0 & u_7 & u_8 - 1 & u_9 \\ u_{10} & 0 & u_{11} & u_{12} - 1 \end{pmatrix}$$

and

$$\det\begin{pmatrix} -1 & \omega^2 u_1 & 0 & \omega^2 u_2 & 0 & 0 & 0 & \omega^2 u_3 \\ \omega u_1 & -1 & \omega u_2 & 0 & 0 & 0 & \omega u_3 & 0 \\ 0 & \omega u_4 & -1 & \omega u_5 & 0 & \omega u_6 & 0 & 0 \\ \omega^2 u_4 & 0 & \omega^2 u_5 & -1 & \omega^2 u_6 & 0 & 0 & 0 \\ 0 & 0 & 0 & \omega^2 u_7 & -1 & \omega^2 u_8 & 0 & \omega^2 u_9 \\ 0 & 0 & \omega u_7 & 0 & \omega u_8 & -1 & \omega u_9 & 0 \\ 0 & \omega u_{10} & 0 & 0 & 0 & \omega u_{11} & -1 & \omega u_{12} \\ \omega^2 u_{10} & 0 & 0 & 0 & \omega^2 u_{11} & 0 & \omega^2 u_{12} & -1 \end{pmatrix}$$

must equal the determinant of the matrix $\widetilde{Z}_{\text{spec}} - I$ given by

$$\begin{pmatrix} u_5 - 1 & u_4 u_1 & u_4 u_2 u_5 & u_4 u_2 u_4 & 0 & u_6 & u_4 u_3 & u_4 u_2 u_6 \\ u_2 & u_1 u_1 - 1 & u_1 u_2 u_5 & u_1 u_2 u_4 & u_3 & 0 & u_1 u_3 & u_1 u_2 u_6 \\ u_6 u_7 & u_4 & u_5 u_5 - 1 & u_5 u_4 & u_6 u_9 & u_6 u_8 & 0 & u_5 u_6 \\ u_3 u_{11} u_7 & u_3 u_{10} & u_2 & u_1 - 1 & u_3 u_{11} u_9 & u_3 u_{11} u_8 & u_3 u_{12} & 0 \\ 0 & u_{10} u_1 & u_{10} u_2 u_5 & u_{10} u_2 u_4 & u_{12} - 1 & u_{11} & u_{10} u_3 & u_{10} u_2 u_6 \\ u_8 u_7 & 0 & u_7 u_5 & u_7 u_4 & u_8 u_9 & u_8 u_8 - 1 & u_9 & u_7 u_6 \\ u_{12} u_{11} u_7 & u_{12} u_{10} & 0 & u_{10} & u_{12} u_{11} u_9 & u_{12} u_{11} u_8 & u_{12} u_{12} - 1 & u_{11} \\ u_9 u_{11} u_7 & u_9 u_{10} & u_7 & 0 & u_9 u_{11} u_9 & u_9 u_{11} u_8 & u_9 u_{12} & u_8 - 1 \end{pmatrix}.$$

21

Non-isomorphic regular graphs without loops or multiedges having the same Ihara zeta function

Algebraic number fields K_1, K_2 can have the same Dedekind zeta functions without being isomorphic. See Perlis [99]. The smallest examples have degree 7 over $\mathbb{Q}$ and come from Artin L-functions of induced representations from subgroups of $G = \mathrm{GL}(3, \mathbb{F}_2)$, the simple group of order 168. An analogous example of two graphs (each having seven vertices) which are isospectral but not isomorphic was given by P. Buser. These graphs are found in Figure 21.1. See Buser [23] or Terras [133], Chapter 22. Buser's graphs ultimately lead to two planar isospectral drums which are not obtained from each other by rotation and/or translation, answering the question raised by M. Kac in [65]: Can you hear the shape of a drum? See Gordon, Webb, and Wolpert [45], who showed using the same basic construction that there are pairs of (non-convex) planar drums that cannot be heard.

Buser's graphs are not simple. That is, they have multiple edges as well as loops. We can use our theory to obtain examples of simple regular graphs with 28 vertices which are isospectral but not isomorphic. See Figure 21.2. The graphs in Figure 21.2 are constructed using the same group G and subgroups H_j as in Buser's examples. Sunada [127] showed how to apply the method from number theory to obtain isospectral compact connected Riemannian manifolds that are not isometric.

Define $G = \mathrm{GL}(3, \mathbb{F}_2)$, which is the group of all non-singular 3×3 matrices with entries in the finite field with two elements. As mentioned above, it is a simple group of order 168. Two subgroups H_j of index 7 in G are

$$H_1 = \left\{ \begin{pmatrix} 1 & 0 & 0 \\ * & * & * \\ * & * & * \end{pmatrix} \right\} \quad \text{and} \quad H_2 = \left\{ \begin{pmatrix} 1 & * & * \\ 0 & * & * \\ 0 & * & * \end{pmatrix} \right\},$$

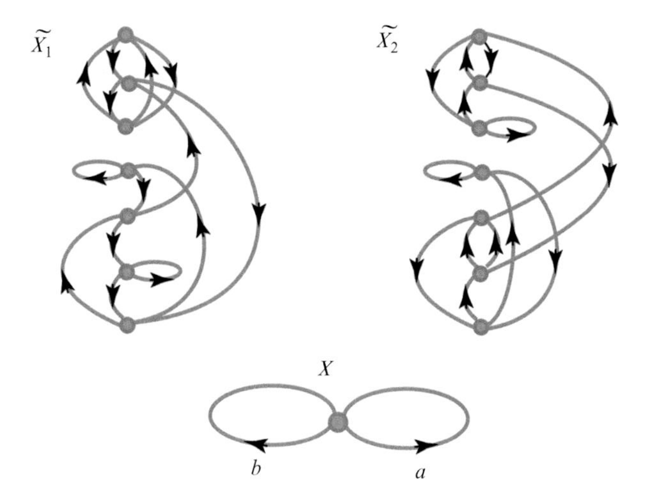

Figure 21.1 Buser's isospectral non-isomorphic Schreier graphs. See Buser [23]. The sheets of $\widetilde{X}_1$ and $\widetilde{X}_2$ are numbered 1 to 7 from bottom to top. The lifts of a are on the right in each graph; the lifts of b are on the left.

where the star means an arbitrary element of the field with two elements, i.e., 0 or 1.

Exercise 21.1 Show that H_1 and H_2 are not conjugate in G.

The preceding exercise also follows from the fact that we can construct two non-isomorphic intermediate graphs corresponding to these subgroups H_j of G. One can show that these two groups give rise to equivalent permutation representations of G (i.e., the representations we have called $\mathrm{Ind}_{H_i}^G 1$). The same argument as was used in Terras [133] for Buser's graphs says that the representations $\rho_j = \mathrm{Ind}_{H_j}^G 1$ are equivalent because the subgroups H_j are almost conjugate (i.e. $|H_1 \cap \{g\}| = |H_2 \cap \{g\}|$ for every conjugacy class $\{g\}$ in G). This implies that we have equality of the corresponding characters, $\chi_{\rho_1} = \chi_{\rho_2}$. Therefore we will get graphs with the same zeta functions (using the induction property of vertex L-functions):

$$\zeta_{\widetilde{X}_1}(u) = L(u, \rho_1) = L(u, \rho_2) = \zeta_{\widetilde{X}_2}(u).$$

See Terras [133] for more information.

Given $g \in G$, all elements of $H_1 g$ have the same first row. The seven possible non-zero first rows correspond naturally to the numbers 1–7 in binary. Thus, order the seven right cosets $H_1 g_j$ by the numbers represented by the first rows in binary. For example, the first row of g_6 is (110) and $H_1 g_4$ is the identity coset. For any g it is easy to work out what coset $H_1 g_j g$ is, as the first

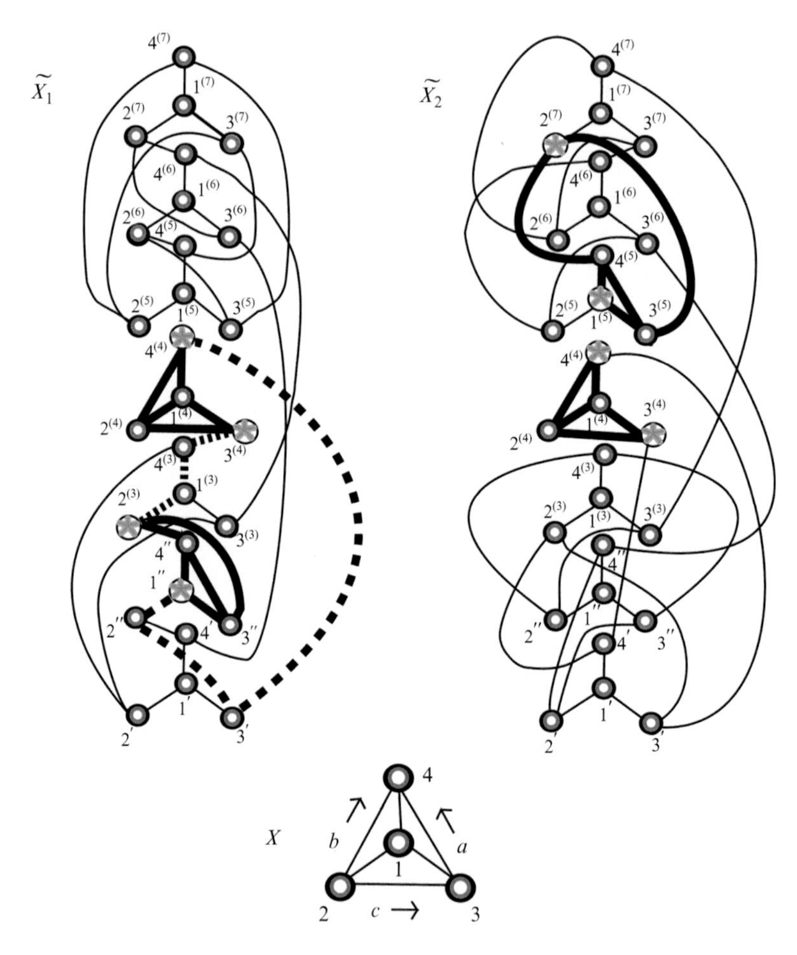

Figure 21.2 Non-isomorphic graphs without loops or multiedges having the same Ihara zeta function. The superscripts number the sheets of $\widetilde{X}_1$ and $\widetilde{X}_2$. The lifts of a are on the right-hand side of each graph and the lifts of b are on the left-hand side; the lifts of c cross from the left to the right.

row of the product $g_j g$ depends only on the first row of g_j. So, for $g \in G$, we can easily find the permutation $\mu(g)$ corresponding to multiplying the seven cosets $H_1 g_j$ by g on the right, i.e., $H_1 g_j g = H_1 g_{\mu(j)}$.

We need the permutations $\mu(A)$ and $\mu(B)$ for Buser's matrices:

$$A = \begin{pmatrix} 0 & 1 & 1 \\ 0 & 1 & 0 \\ 1 & 0 & 0 \end{pmatrix} \quad \text{and} \quad B = \begin{pmatrix} 1 & 0 & 0 \\ 0 & 0 & 1 \\ 0 & 1 & 1 \end{pmatrix}. \tag{21.1}$$

Computation shows that

$$\mu(A) = (1436)(2)(57) \qquad \text{and} \qquad \mu(B) = (132)(4)(576). \tag{21.2}$$

Exercise 21.2 Check these formulas. For example, to find $H_1 g_3 A$, we want the first row of

$$\begin{pmatrix} 0 & 1 & 1 \\ * & * & * \\ * & * & * \end{pmatrix} \begin{pmatrix} 0 & 1 & 1 \\ 0 & 1 & 0 \\ 1 & 0 & 0 \end{pmatrix} = \begin{pmatrix} 1 & 1 & 0 \\ * & * & * \\ * & * & * \end{pmatrix} \in H_1 g_6$$

and so $\mu(A)$ takes 3 to 6.

We now need to do the same permutation calculation with the matrices A and B acting on the right cosets of H_2. It might appear that the right cosets of H_2 would be more difficult to deal with. But there is a very useful automorphism of G to help. It is $\varphi(g) = {}^{\mathrm{t}}g^{-1}$, where ${}^{\mathrm{t}}g$ denotes the transpose of $g \in G$. This map φ is an automorphism of G such that $\varphi(H_1) = H_2$. If we apply φ to the right cosets $H_1 g_j$, we get G as a union of the seven right cosets $H_2\, {}^{t}g_j^{-1}$. To work out how $g \in G$ permutes the cosets, it suffices to consider the action of ${}^{\mathrm{t}}g^{-1}$ on the $H_1 g_j$. Note that

$${}^{\mathrm{t}}A^{-1} = \begin{pmatrix} 0 & 0 & 1 \\ 0 & 1 & 1 \\ 1 & 0 & 0 \end{pmatrix} \qquad \text{and} \qquad {}^{\mathrm{t}}B^{-1} = \begin{pmatrix} 1 & 0 & 0 \\ 0 & 1 & 1 \\ 0 & 1 & 0 \end{pmatrix}.$$

Therefore the action of ${}^{t}A^{-1}$ and ${}^{t}B^{-1}$ on the right cosets $H_1 g_j$ is given by the permutations

$$\mu({}^{\mathrm{t}}A^{-1}) = (14)(2376)(5) \quad \text{and} \quad \mu({}^{\mathrm{t}}B^{-1}) = (123)(4)(567). \tag{21.3}$$

These same permutations give the actions of A and B on the right cosets $H_2\, {}^{\mathrm{t}}g_j^{-1}$.

Exercise 21.3 Check these formulas for $\mu({}^{t}A^{-1})$ and $\mu({}^{t}B^{-1})$.

Exercise 21.4 Prove that the matrices A and B in formula (21.1) generate the group G.

Buser [23] used the matrices A and B to construct two Schreier graphs corresponding to the two subgroups H_1 and H_2. Using the Galois theory we worked out in the preceding chapters, this means find coverings $\widetilde{X_1}$ and $\widetilde{X_2}$ of X, where X is the graph consisting of a single vertex and a double loop. Direct each loop, so that we have two directed edges, a and b, say. Assign the normalized Frobenius elements $\sigma(a) = A$ and $\sigma(b) = B$. The resulting normal cover of X is the Cayley graph of G corresponding to the generators A and B. We want two intermediate graphs $\widetilde{X_1}$ and $\widetilde{X_2}$ that correspond to the subgroups H_1 and H_2 in accordance with Theorem 14.3; these are Schreier graphs. The permutations $\mu(A)$ and $\mu(B)$ that we have just found tell us how to lift the edges a and b. This tells us how to draw the graphs $\widetilde{X_1}$ and $\widetilde{X_2}$. See Figure 21.1.

There are many ways to see that the two graphs in Figure 21.1 are not isomorphic – even as undirected graphs. Look at the triple edges; look at the double edges; look at the distances between loops; etc. Therefore H_1 and H_2 are not conjugate in G. Both graphs are 4-regular; they have the same zeta function and their adjacency matrices have the same spectrum.

Next we are going to **construct** graphs like Buser's that have no loops or multiple edges i.e., **simple, 3-regular, isospectral non-isomorphic graphs**. We use the same G, H_1, and H_2 but take X to be a tetrahedron K_4. Thus X is 3-regular and has a fundamental group of rank 3. Take the cut or deleted edges (directed as in Figure 21.2) to be a, b, c. Choose the normalized Frobenius automorphisms to be

$$\sigma(a) = A, \qquad \sigma(b) = \sigma(c) = B.$$

Take seven copies of the tree of X to be the sheets of $\widetilde{X_1}$ and again for $\widetilde{X_2}$. On $\widetilde{X_1}$ we lift a, b, c, using the permutations $\mu(A)$ and $\mu(B)$ from formula (21.2) above. On $\widetilde{X_2}$ we lift a, b, c using the permutations $\mu({}^tA^{-1})$ and $\mu({}^tB^{-1})$ from formula (21.3). This produces the graphs $\widetilde{X_1}$ and $\widetilde{X_2}$ shown in Figure 21.2.

Both graphs are 3-regular; they have the same zeta function and their adjacency matrices have the same spectrum. The proof that

$$\zeta_{\widetilde{X_1}}(u) = L(u, \rho_1) = L(u, \rho_2) = \zeta_{\widetilde{X_2}}(u)$$

employs the same argument as that used above and in Terras [133] for Buser's graphs.

Let us say a bit more about the construction that leads to Figure 21.2. The edge c goes from vertex 2 to vertex 3 in X and has the normalized Frobenius automorphism $\sigma(c) = B$. The lifts of c to $\widetilde{X_1}$ are determined by the permutation $\mu(B) = (132)(4)(576)$ from (21.2). This means that c in X lifts to an edge in $\widetilde{X_1}$ from $2'$ to $3^{(3)}$, an edge from $2^{(3)}$ to $3^{(2)}$, an edge from

$2''$ to $3'$, and then (beginning a new cycle) to an edge from $2^{(4)}$ to $3^{(4)}$, etc. The edge b lifts in exactly the same manner as c. Similarly, for $\widetilde{X_1}$, the edge a in X corresponds to the permutation $(1436)(2)(57)$ from (21.2). This means that edge a in X lifts to an edge in $\widetilde{X_1}$ from $3'$ to $4^{(4)}$, an edge from $3^{(4)}$ to $4^{(3)}$, an edge from $3^{(3)}$ to $4^{(6)}$, etc. To get $\widetilde{X}_2$ proceed similarly, using the permutations in (21.3).

To see that graphs $\widetilde{X_1}$ and $\widetilde{X_2}$ in Figure 21.2 are not isomorphic, we proceed as follows. There are exactly four triangles in each graph (shown by the thick solid lines) and they are connected in pairs in both graphs. This distinguishes in each pair the two vertices not on common edges (the vertices containing stars). In $\widetilde{X_1}$ we can go in three steps (via the broken lines) from a star vertex in one pair to a star vertex in the other pair, in two different ways, one indicated by the lines with short breaks and the other by the lines with longer breaks. This cannot be done at all in $\widetilde{X_2}$.

We said that each $\widetilde{X_i}$ has four triangles (up to equivalence and choice of direction). Why? Since X has no loops or multiedges, any triangle on $\widetilde{X_1}$ or $\widetilde{X_2}$ projects to a triangle on X. We saw back in Chapter 1 that (up to equivalence and choice of direction) $X = K_4$ has eight primes of length 3 and therefore four triangles.

Let χ_1 be the trivial character on H_1 or H_2. The induced representations $\mathrm{Ind}_{H_i}^G 1$ for $i = 1, 2$, have the same character $\chi_1^{\#}$. By Corollary 19.11, for any directed triangle C on X there are $\chi_1^{\#}(\sigma(C))$ directed triangles above C on $\widetilde{X_1}$ and also above C on $\widetilde{X_2}$. Reversing the direction of C reverses the direction of the covering triangles. We will choose the most convenient direction for each triangle.

Three triangles on X have two edges on the tree of X with normalized Frobenius elements equal to 1 automatically. Thus, with an appropriate choice of direction in each case, $\sigma(C) = A, B, B$, for each triangle. The fourth triangle may be taken to be the path $ab^{-1}c$ whose normalized Frobenius automorphism is $\sigma(a)\sigma(b)^{-1}\sigma(c) = AB^{-1}B = A$. For $g \in G$, $\chi_1^{\#}(g)$ is simply the number of 1-cycles in the permutation $\mu(g)$. In particular, $\chi_1^{\#}(A) = \chi_1^{\#}(B) = 1$ (the same for both H_1 and H_2). Thus each of the four triangles of X has precisely one triangle of $\widetilde{X_j}$ above it for $j = 1$ or 2. Thus the triangles shown in Figure 21.2 are all the triangles on $\widetilde{X_1}$ and $\widetilde{X_2}$, as we claimed.

There are many other results on isospectral graphs. For infinite towers of isospectral graphs coming from finite simple groups, see Alexander Lubotzky, Beth Samuels, and Uzi Vishne [80]. Many examples (not necessarily connected) have been found by an undergraduate research group (REU) at Louisiana State University directed by Robert Perlis. See the papers by Rachel

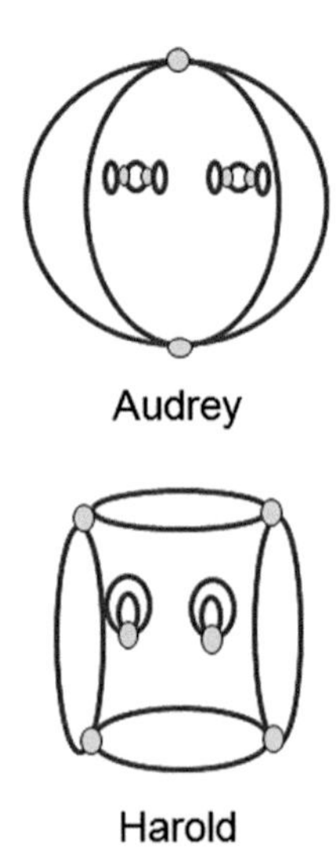

Figure 21.3 Two isospectral non-isomorphic graphs, named Harold and Audrey found by students in the Research Experiences for Undergraduates program at Louisana State University and drawn for me by Aubi Mellein. There is a different version of the picture in Rachel Reeds' paper [103].

Reeds [103] and Yaim Cooper [24] for some results. We include the Harold and Audrey graphs found by this REU in Figure 21.3. They were shown to me by Aubi Mellein, who had taken part in the REU. Storm [124] shows that isospectral irregular graphs need not have the same Ihara zeta function.

Other questions can be asked. See Ram Band, Talia Shapira, and Uzy Smilansky [10] for the question whether a count of nodal domains for the eigenvectors resolves the isospectrality of our examples. This question is "Can one count the shape of a drum?" Classically a nodal domain is a maximally connected region where the eigenfunction ψ of $-\Delta$ for a bounded region $D \subset \mathbb{R}^n$ (ψ satisfying Dirichlet boundary conditions on ∂D, meaning that it vanishes on ∂D, or Neumann boundary conditions, meaning that its normal derivative vanishes on ∂D) has a constant sign. If $n = 1$, Sturm's oscillation theorem states that the nth eigenfunction has exactly n nodal domains. Here the eigenfunctions are ordered by increasing eigenvalues. In higher dimensions Courant showed that the number of nodal domains of the nth eigenfunction is less than or equal to n.

In [10] the conjecture is stated that nodal counts resolve the isospectrality of isospectral quantum graphs; quantum graphs are weighted graphs whose Schödinger operator on an edge is the one-dimensional Laplacian. There are boundary conditions (say Neumann) at the vertices. A **wavefunction** is a function on each edge that is continuous at the vertices and satisfies boundary conditions. Let S_i be the set of edges from vertex i. The wavefunction ψ_b with wave number k can be written, for vertex i and vertex j connected by edge b of length L_b and with coordinate x_b along the edge, as follows:

$$\psi_b(x_b) = \frac{1}{\sin(kL_b)} \left\{ \phi_i \sin(k(L_b - x_b)) + \phi_j \sin(kx_b) \right\},$$

$$\sum_{b \in S_i} \frac{d}{dz_b} \psi_b(x_b) \bigg|_{x_b=0} = 0.$$

Here the wavefunction ψ_b takes the values ϕ_i and ϕ_j at vertices i and j, respectively. Substitute the first equation into the second and obtain equations for the ϕ_j:

$$\sum_{j=1}^{|E|} A_{i,j} \left(L_1, \ldots, L_{|E|}; k\right) \phi_j = 0 \qquad \text{for all } 1 \le i \le |V|.$$

The spectrum $\{k_n\}$ is a discrete, positive, unbounded sequence, the zero set of the determinant of the matrix of coefficients $A_{i,j} \left(L_1, \ldots, L_{|E|}; k\right)$. Then one must regularize the determinant function.

There are two ways to define the nodal domains for quantum graphs. The discrete way says that a nodal domain is a maximal set of connected interior vertices (meaning vertices with degree ≥ 3) where the vertex eigenfunctions ϕ_i have the same sign. This definition is modified if any ϕ_i vanishes.

It has been shown that isospectral pairs of quantum graphs must have rationally dependent edge lengths.

22
Chebotarev density theorem

The Chebotarev density theorem for algebraic number fields was proved in 1922. There are discussions in Stevenhagen and Lenstra [123] and Stark [118]. The Stark version is summarised in Figure 22.1. The Chebotarev density theorem generalizes the Dirichlet theorem, saying that there is an infinite number of primes in an arithmetic progression of the form $\{a + nb | n \in \mathbb{Z}\}$, when a, b are relatively prime. It also generalizes a theorem of Frobenius from 1880 concerning a monic irreducible polynomial $f(x) \in \mathbb{Z}[x]$ of degree n, with non-zero discriminant $\Delta(f)$, and the list of degrees $e_1, \ldots, e_t$ of the irreducible factors mod p (called the **decomposition type** mod p) for primes p not dividing $\Delta(f)$. The decomposition type of $f(\text{mod } p)$ is a partition of n; i.e., $n = e_1 + \cdots + e_t$. Let K be the extension field of $\mathbb{Q}$ obtained by adjoining all the roots of $f(x)$ and let G be the Galois group of $K/\mathbb{Q}$. The **Frobenius density theorem** says that the density (from Figure 22.1) of such p for a given decomposition type $e_1, \ldots, e_t$ is

$$\frac{\#\{\sigma \in G | \text{cycle pattern of } \sigma \text{ is } e_1, \ldots, e_t\}}{|G|}. \tag{22.1}$$

Here by a **cycle pattern** we mean that the Galois group is viewed as a subgroup of the symmetric group of permutations of the roots of the polynomial $f(x)$. Then σ is written as a product of disjoint cycles (including cycles of length 1). The cycle pattern is the list of cycle lengths. It is also a partition of n.

In order to prove the graph theory analog of the Chebotarev density theorem we will need some information on the poles of zeta and L-functions of graph coverings. We begin with a lemma for which you need to recall our notation for the largest circle of convergence R_X from Definition 2.4 and the edge adjacency matrix $W_1 = W_1(X)$ from Definition 4.1. Recall that R_X denotes the

Chebotarev density theorem for $K/\mathbb{Q}$ normal.
For a set S of rational primes, define the analytic density of S to be $\lim_{s\to 1+} \frac{\sum_{p\in S} p^{-s}}{\log[1/(s-1)]}$

In the follwing proof, one needs to know that $L(s,\pi)$ continues to $s=1$ with no pole or zero if $\pi \neq 1$, while $L(s,1) = \zeta(s)$ is the Riemann zeta.

Theorem Define $C(p)$ as the conjugacy class of the Frobenius automorphism of prime ideals $\mathfrak{P}$ of K above p. Then $\forall$ a conjugacy class C in $G=\mathrm{Gal}(K/\mathbb{Q})$, the analytic density of the set of rational primes p such that $C=C(p)$ is $|C|/|G|$.

Proof Sum the logs of the Artin L-functions times the conjugates of the corresponding characters χ_π over all irreducible reps π of G. As $s \to 1+$, we have

$$\log\frac{1}{s-1} \sim \sum_{\pi} (\log L(s,\pi))\,\overline{\chi_\pi(C)}$$

$$\sim \sum_{\pi}\sum_{p} \chi_\pi(C(p))\,p^{-s}\,\overline{\chi_\pi(C)} \sim \frac{|G|}{|C|}\sum_{\substack{p \\ C(p)=C}} p^{-s},$$

by the orthogonality ralations of the characters of the irreducible representations π of G. □

Figure 22.1 The Chebotarev density theorem in the number field case, i.e., for $K/\mathbb{Q}$ normal.

radius of convergence of the Ihara zeta function of X as well as the closest pole of zeta to 0 and that R_X is the reciprocal of the Perron–Frobenius eigenvalue of W_1.

Lemma 22.1 *Suppose that Y is an n-sheeted covering of X. The maximal absolute value of an eigenvalue of the edge adjacency matrix $W_1(X)$ is the same as that for $W_1(Y)$. This common value is $R_Y^{-1} = R_X^{-1}$.*

Proof First recall that $\zeta_X^{-1}(u)$ divides $\zeta_Y^{-1}(u)$. See Proposition 13.10. It follows that $R_Y \leq R_X$.

Then a standard estimate from the theory of zeta functions of number fields works for graph theory zeta functions as well. For all real $u \geq 0$ such that the infinite product for $\zeta_X(u)$ converges, we have

$$\zeta_Y(u) \leq \zeta_X(u)^n, \tag{22.2}$$

with n equal to the number of sheets of the covering. Thus $R_X \leq R_Y$.

We will take the method of proof of formula (22.2) from Lang [73], p. 160. One begins with the product formula for $\zeta_Y(u)$ and the behavior of primes in coverings.

So, for real u such that $R_Y > u \geq 0$ we have a product over primes $[D]$ of Y, giving

$$\zeta_Y(u) = \prod_{[D]} \left(1 - u^{\nu(D)}\right)^{-1}.$$

Rewrite this as a product over primes $[C]$ of X and then primes of Y $[D_1], \ldots, [D_g]$, all above $[C]$:

$$\zeta_Y(u) = \prod_{[C]} \prod_{i=1}^{g} \left(1 - u^{\nu(D_i)}\right)^{-1}.$$

Recall that the prime D_i above C is a closed path obtained by lifting C a total of f_i times. This means that

$$\zeta_Y(u) = \prod_{[C]} \prod_{i=1}^{g} \left(1 - u^{f_i \nu(C)}\right)^{-1}.$$

We know (from Exercise 15.2 on the behavior of primes in non-normal coverings) that $n = \sum_{i=1}^{g} f_i$. It follows that

$$\zeta_Y(u) \leq \prod_{[C]} \left(1 - u^{\nu(C)}\right)^{-g} \leq \zeta_X(u)^n.$$

□

Next let us define the analytic density.

Definition 22.2 If S is a set of primes in X, we define the **analytic density**

$$\delta(S) = \lim_{u \to R_X -} \frac{\sum_{[C] \in S} u^{\nu(C)}}{\sum_{[C]} u^{\nu(C)}} = \lim_{u \to R_X -} \frac{\sum_{[C] \in S} u^{\nu(C)}}{\log \zeta_X(u)} = \lim_{u \to R_X -} \frac{\sum_{[C] \in S} u^{\nu(C)}}{- \log(R_X - u)}.$$

Here the sums are over primes $[C]$ in X.

Question Why does $\sum_{[C]} u^{\nu(C)}$ blow up at $u = R_X$ in the same way as $\log \zeta_X(u)$?

To answer this, recall that, for $0 \le u < R_X$,

$$\zeta_X(u) = \prod_{[C]} \left(1 - u^{\nu(C)}\right)^{-1} = \det(I - uW_1(X))^{-1}.$$

Take the logarithm and obtain

$$\log \zeta_X(u) = -\sum_{[C]} \log\left(1 - u^{\nu(C)}\right) = \sum_{[C]} \sum_{m \ge 1} \frac{1}{m} u^{m\nu(C)}$$
$$= \sum_{[C]} u^{\nu(C)} + H(u),$$

where

$$H(u) = \sum_{[C]} \sum_{m \ge 2} \frac{1}{m} u^{m\nu(C)}.$$

The amazing thing is that $H(u)$ is bounded up to $u = R_X$ and beyond. Thus $\sum_{[C]} u^{\nu(C)}$ must account for the blowup of $\log \zeta_X(u)$ at $u = R_X$.

To see that $H(u)$ is bounded up to $u = R_X$, we need to make a few estimates:

$$\sum_{m \ge 2} \frac{1}{m} u^{m\nu(C)} \le \sum_{m \ge 2} u^{m\nu(C)} = \frac{u^{2\nu(C)}}{1 - u^{\nu(C)}} \le \frac{u^{2\nu(C)}}{1 - u}.$$

Here we are using the fact that $0 \le u \le R_X \le 1$. It follows that

$$H(u) \le \frac{1}{1-u} \sum_{[C]} u^{2\nu(C)},$$

which converges up to $u^2 = R_X$ or $u = \sqrt{R_X}$ as $\log \zeta_X(u^2)$ converges up to $u = \sqrt{R_X}$.

What if $R_X = 1$? Then the graph must be a cycle. Why? The reason is that $p = q = 1$, using results from Chapter 8.

You might prefer to use a less complicated notion of density than that in Definition 22.2. Perhaps you would like to say that a set S of primes has a **natural density** δ if

$$\frac{\{[P] | [P] = \text{prime},\ [P] \in S,\ \text{and } \nu(P) \le n\}}{\{[P] | [P] = \text{prime and } \nu(P) \le n\}} \to \delta \qquad \text{as } n \to \infty.$$

At least in the number theory case, the proof of the density theorem is harder with this version of the density. For rational primes, if a set of primes has a natural density then it has an analytic density and the two densities are equal. However, the converse is false. We leave it as a **research problem** to work out what happens in the graph theory case.

Theorem 22.3 (Graph theory Chebotarev density theorem) *Suppose that the graph X satisfies our usual hypotheses. If Y/X is normal and $\{g\}$ is a fixed conjugacy class in the Galois group $G = G(Y/X)$ then*

$$\delta\{[C] \text{ prime of } X \mid \sigma(C) = \{g\}\} = \frac{|\{g\}|}{|G|}.$$

Here $\sigma(C)$ is the normalized Frobenius automorphism for C.

Proof We will imitate the proof sketched by Stark in [118] for the number field case, where one knows much less about the Artin L-functions. The idea goes back to Dirichlet. The main idea is to sum the terms $\overline{\chi_\pi(g)} \log L(u, \pi, Y/X)$ over all irreducible representations π of G. Here $\chi_\pi = \text{Tr}\, \pi$. This gives the following asymptotic formula as u approaches R_X from below:

$$\log\left(\frac{1}{R_X - u}\right)_{u \to R_X-} \sim \log \zeta_X(u) \sim \sum_{\pi \in \widehat{G}} \overline{\chi_\pi(g)} \log L(s, \pi).$$

Here we have used Lemma 22.1 and the fact that

$$\zeta_Y(u) = \prod_{\rho \in \widehat{G}} L(u, \rho, Y/X)^{d_\rho}.$$

It follows from the Euler product definition of the L-functions that

$$\log\left(\frac{1}{R_X - u}\right)_{u \to R_X-} \sim \sum_{\pi \in \widehat{G}} \sum_{\substack{[C] \\ \text{prime of } X}} \chi_\pi(\sigma(C)) u^{\nu(C)} \overline{\chi_\pi(g)}$$
$$+ \sum_{\pi \in \widehat{G}} \sum_{\substack{[C] \\ \text{prime of } X}} \sum_{m \geq 2} \frac{1}{m} \chi_\pi(\sigma(C^m)) u^{m\nu(C)} \overline{\chi_\pi(g)}.$$

The second term in the sum is holomorphic as $u \to R_X-$. To see this, note that the second term can be written as

$$\sum_{\substack{[C] \\ \text{prime of } X}} \sum_{m \geq 2} \frac{1}{m} u^{m\nu(C)} \sum_{\pi \in \widehat{G}} \chi_\pi(\sigma(C^m)) \overline{\chi_\pi(g)}.$$

Then, using the orthogonality relations for characters of G dual to those in formula (18.3), we find that this last sum is, for $0 \le u < R_X$,

$$\frac{|G|}{|\{g\}|} \sum_{\substack{[C] \\ \{\sigma(C^m)\}=\{g\}}} \sum_{m \ge 2} \frac{1}{m} u^{m\nu(C)} \le |G| \sum_{[C]} \sum_{m \ge 2} \frac{1}{m} u^{m\nu(C)}$$

$$\le |G| \sum_{[C]} \frac{u^{2\nu(C)}}{1-u} \le \frac{|G|}{1-u} \sum_{[C]} u^{2\nu(C)}.$$

This is holomorphic up to $u = \sqrt{R_X}$.

Thus, we have shown that $\log(1/(R_X - u))$ is asymptotic to the following as $u \to R_X-$:

$$\sum_{\pi \in \widehat{G}} \sum_{\substack{[C] \\ \text{prime of } X}} \chi_\pi(\sigma(C)) u^{\nu(C)} \overline{\chi_\pi(g)} = \sum_{\substack{[C] \\ \text{prime of } X}} \sum_{\pi \in \widehat{G}} \chi_\pi(\sigma(C)) u^{\nu(C)} \overline{\chi_\pi(g)}$$

$$= \frac{|G|}{|\{g\}|} \sum_{\substack{[C] \\ \{\sigma(C)\}=\{g\}}} u^{\nu(C)}.$$

For the last equality, use the dual orthogonality relations again. The theorem follows. □

The simplest example to which we can apply the Chebotarev theorem is the cube over the tetrahedron, where primes with $f = 1$ have density 1/2 as do the primes with $f = 2$. See Figures 15.2 and 15.3.

A more complicated example of our result can be found in Figure 15.4. This example concerns the splitting of primes in a non-normal cubic covering Y_3/X, where $X = K_4 - e$. Thus one must consider what happens in the normal cover for which Y_3/X is intermediate.

Exercise 22.1 Fill in the details concerning the densities of the primes in various classes in Figure 15.4 by imitating Stark's arguments for the corresponding example in [118], pp. 358–360 and 364. The Frobenius version of Chebotarev's theorem in formula (22.1) would simplify the computation.

23

Siegel poles

23.1 Summary of Siegel pole results

In number theory there is a known zero-free region for a Dedekind zeta function, which can be given explicitly except for the possibility of a single first-order real zero within this region. This possible exceptional zero has come to be known as a **Siegel zero** and is closely connected with the Brauer–Siegel theorem on the growth of the class number times the regulator with the discriminant. See Lang [73] for more information on the implications of the non-existence of Siegel zeros. There is no known example of a Siegel zero for Dedekind zeta functions. In number fields, a Siegel zero (should it exist) "deserves" to arise already in a quadratic extension of the base field. This has now been proved in many cases (see Stark [116]).

The reciprocal of the Ihara zeta function, $\zeta_X(u)^{-1}$, is a polynomial with a finite number of zeros. Thus there is an $\epsilon > 0$ such that any pole of $\zeta_X(u)$ in the region $R_X \leq |u| < R_X + \epsilon$ must lie on the circle $|u| = R_X$. This gives us the graph theoretic analog of a **pole-free region**, $|u| < R_X + \epsilon$; the only exceptions lie on the circle $|u| = R_X$. We will show that $\zeta_X(u)$ is a function of u^δ with $\delta = \delta_X$ a positive integer from Definition 23.2 below. This implies that there is a δ-fold symmetry in the poles of $\zeta_X(u)$, i.e., that $u = \varepsilon_\delta R$ is also a pole of $\zeta_X(u)$ for all δth roots of unity ε_δ. Any further poles of $\zeta_X(u)$ on $|u| = R$ will be called **Siegel poles** of $\zeta_X(u)$. Thus if $\delta = 1$ then any pole $u \neq R$ of $\zeta_X(u)$ with $|u| = R$ will be called a Siegel pole.

All graphs considered in this chapter are assumed to satisfy our usual hypotheses stated in Section 2.1.

Definition 23.1 A vertex of X having degree ≥ 3 is called a **node** of X.

A graph X satisfying our usual hypotheses will have rank ≥ 2 and thus always has at least one node.

Now the reader should recall Definition 2.12, the definition of Δ. We will consider next a closely related quantity.

Definition 23.2 Define δ to be

$$\delta_X = \text{g.c.d.} \left\{ \nu(P) \,\middle|\, \begin{array}{l} P = \text{backtrackless path in } X \text{ such that the} \\ \text{initial and terminal vertices are both nodes} \end{array} \right\}.$$

When a path P in the definition of δ_X is closed, the path will be backtrackless but may have a tail. Later we give an equivalent definition of δ_X not involving paths with possible tails. The relation between δ_X and our earlier Δ_X from Definition 2.12 is given by the following result. As a consequence we will see that any k-regular graph X with $k \geq 3$ has $\delta_X = 1$.

Theorem 23.3 *Suppose that the graph X satisfies our usual hypotheses. Then either* $\Delta_X = \delta_X$ *or* $\Delta_X = 2\delta_X$.

It is easy to see that if Y is a covering graph of X (of rank ≥ 2) we have $\delta_Y = \delta_X$ since they are the g.c.d.s of the same set of numbers. Therefore δ_X is a covering invariant. Because of this, Theorem 23.3 has a useful corollary.

Corollary 23.4 *If Y is a covering of a graph X of rank* ≥ 2 *then*

$$\Delta_Y = \Delta_X \text{ or } 2\Delta_X.$$

For a cycle graph X the ratio Δ_Y/Δ_X can be arbitrarily large. The general case of the theorem about Siegel poles can be reduced to the more easily stated case $\delta_X = 1$, for which any pole of $\zeta_X(u)$ on $|u| = R$ other than $u = R$ is a **Siegel pole**.

Theorem 23.5 (Siegel poles when $\delta_X = 1$) *Suppose that X satisfies our usual hypotheses and that* $\delta_X = 1$. *Let Y be a covering graph of X and suppose that* $\zeta_Y(u)$ *has a Siegel pole* μ. *Then we have the following facts.*

(1) *The pole* μ *is a first-order pole of* $\zeta_Y(u)$ *and* $\mu = -R$ *is real.*
(2) *There is a unique graph* X_2 *intermediate to* Y/X *with the property that, for every graph* $\widetilde{X}$ *intermediate to* Y/X *(including* X_2*),* μ *is a Siegel pole of* $\zeta_{\widetilde{X}}(u)$ *if and only if* $\widetilde{X}$ *is intermediate to* Y/X_2.
(3) *The graph* X_2 *is either X or a quadratic (i.e., 2-sheeted) cover of X.*

23.2 Proof of Theorems 23.3 and 23.5

Before proving Theorem 23.3, we need a lemma.

Lemma 23.6 *The invariant δ of Definition 23.2 equals the invariant δ' defined by*

$$\delta' = \text{g.c.d.}\left\{ \nu(P) \,\middle|\, \begin{array}{l} P \text{ is backtrackless, the initial and terminal} \\ \text{vertices of } P \text{ are (possibly equal) nodes,} \\ \text{no intermediate vertex is a node} \end{array} \right\}.$$

Proof Clearly $\delta|\delta'$, i.e., δ divides δ'.

To show that $\delta'|\delta$, note that anything in the length set for δ is a sum of elements of the length set for δ'. □

Exercise 23.1 Use Lemma 23.6 to show that any k-regular graph X with $k \geq 3$ has $\delta_X = 1$.

Exercise 23.2 Compute Δ_X and δ_X for K_4, the cube, and for $K_4 - e$, where e is an edge.

Proof of Theorem 23.3 Theorem 23.3 says that if Δ_X is odd then $\Delta_X = \delta_X$; otherwise either $\Delta_X = \delta_X$ or $\Delta_X = 2\delta_X$. First note that $\delta|\Delta$ since every cycle in a graph X of rank ≥ 2 has a node (otherwise X would not be connected). To finish the proof, we show that $\Delta|2\delta$.

If X has a loop then the vertex of the loop is a node (if the rank is ≥ 2) and thus $\Delta = \delta = 1$. **So assume that X is loopless for the rest of the proof.**

By Lemma 23.6 we may consider only backtrackless paths A between arbitrary nodes α_1 and α_2 without intermediate nodes. There are two cases.

Case 1 $\alpha_1 \neq \alpha_2$.

Let e_1' be an edge out of α_1 not equal to the initial edge i of A (or therefore to i^{-1} since there are no loops). Let e_2' be an edge into α_2 not equal to the terminal edge t of A (or to t^{-1}). Let $B = P\left(e_1', e_2'\right)$ from Lemma 11.11.

Suppose that e_1'' is another edge out of α_1 such that $e_1'' \neq i$, $e_1'' \neq e_1'$ (or their inverses). Likewise suppose e_2'' is another edge into α_2 such that $e_2'' \neq t$, $e_2'' \neq e_2'$ (or their inverses). Let $C = P\left(e_1'', e_2''\right)$ from Lemma 11.11. See Figure 23.1. Then AB^{-1}, AC^{-1}, BC^{-1} are backtrackless tailless paths from α_1 to α_1.

We have

$$\Delta|\nu(AB^{-1}) = \nu(A) + \nu(B),$$

$$\Delta|\nu(AC^{-1}) = \nu(A) + \nu(C),$$

$$\Delta|\nu(BC^{-1}) = \nu(B) + \nu(C).$$

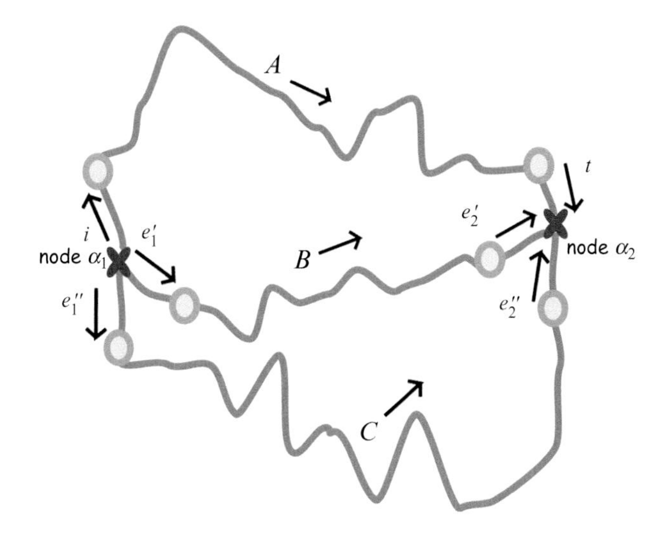

Figure 23.1 The paths in the proof of Theorem 23.3 for the case when the nodes are different.

It follows that Δ divides $2\nu(A)$ since

$$2\nu(A) = (\nu(A) + \nu(B)) + (\nu(A) + \nu(C)) - (\nu(B) + \nu(C)) .$$

Case 2 $\alpha_1 = \alpha_2$.

Then A is a backtrackless path from α_1 to α_1 without intermediate nodes. This implies that A has no tail, since if it did then the other end of the tail would have to be an intermediate node. Therefore Δ divides $\nu(A)$ and hence Δ divides $2\nu(A)$.

Thus, in all cases Δ divides $2\nu(A)$ and hence it divides 2δ, also. □

Lemma 23.7 *We have $\zeta_Y(u) = f(u^d)$ if and only if d divides Δ_Y.*

Proof By definition, $\zeta_Y(u)$ is a function of u^{Δ_Y} and therefore of u^d for all divisors d of Δ_Y.

Conversely, suppose that $\zeta_Y(u)$ is a function of u^d. In the power series $\zeta_Y(u) = \sum_{n \geq 0} a_n u^n$, d divides n for all n for which $a_n > 0$. But if P is a prime cycle of Y with length $\nu(P) = n$ then $a_n \geq 1$ and hence, for all prime cycles P, d divides $\nu(P)$. Therefore, by Definition 2.12, the definition of Δ, we see that d divides Δ_Y. □

Proof of Theorem 23.5 First we need to reduce the theorem to the case where Y/X is normal with Galois group G. To see that this is possible, let $\widetilde{Y}$ be a normal cover of X containing Y. Then $\zeta_{\widetilde{Y}}(u)^{-1}$ is divisible by $\zeta_Y(u)^{-1}$ and both graphs have the same R, by Lemma 22.1, as well as the same δ, which is assumed to be 1. Therefore a Siegel pole of $\zeta_Y(u)$ is a Siegel pole of $\zeta_{\widetilde{Y}}(u)$. Once the theorem is proved for normal covers of X, the graph X_2 which we will have obtained will be contained in Y as well as in every graph intermediate to Y/X whose zeta function has the Siegel pole, and thus we will be done. From now on, assume that Y/X is normal.

Recall Corollary 11.5, i.e., $\zeta_X(u)^{-1} = \det(I - W_X u)$, and Definition 4.1 of the edge adjacency matrix $W_1 = W_X$. The poles of $\zeta_X(u)$ are reciprocal eigenvalues of W_X. For graphs satisfying our usual hypotheses the edge adjacency matrix W_X satisfies the hypothesis of the Perron–Frobenius theorem, namely that W_X is irreducible.

Lemma 22.1 and Theorem 11.16 of Perron and Frobenius imply that if there are d poles of $\zeta_Y(u)$ on $|u| = R_Y = R_X$ then these poles are equally spaced first-order poles on the circle and, further, that $\zeta_Y(u)$ is a function of u^d. Lemma 23.7 implies that Δ_Y has to be divisible by d. But $\delta = \delta_X = \delta_Y = 1$ implies that $\Delta_Y = 1$ or 2. Therefore $d = 1$ or 2. If there is a Siegel pole, $d > 1$. Thus if there is a Siegel pole, $d = 2$, $\Delta_Y = 2$, and the equal-spacing result says that the Siegel pole is $-R_X$ and that it is a pole of order 1.

Corollary 18.11 states that

$$\zeta_Y(u) = \prod_{\pi \in \widehat{G}} L(u, \pi)^{d_\pi}. \tag{23.1}$$

Therefore $L(u, \pi)$ has a pole at $-R_X$ for some π and $d_\pi = 1$. Moreover π must be real or $L(u, \overline{\pi})$ would also have a pole at $-R_X$.

Thus either π is trivial or it is first degree and $\pi^2 = 1$, $\pi \neq 1$, in which case we say that π **is quadratic**.

Case 1 π **is trivial**. Then $\Delta_X = 2 = \Delta_Y$. Every intermediate graph has poles at $-R_X$ as well.

Case 2 $\pi = \pi_2$ **is quadratic**. Then no other $L(u, \pi)$ has $-R_X$ as a pole since it is a first-order pole of $\zeta_Y(u)$. Let

$$H_2 = \{x \in G | \pi_2(x) = 1\} = \ker \pi_2.$$

Then $|G/H_2| = 2$, which implies that there is a graph X_2 corresponding to H_2 by Galois theory. Moreover X_2 is a quadratic cover of X.

Consider the diagram of covering graphs in Figure 23.2; the Galois groups are indicated next to the covering lines. Then

$$\zeta_{\widetilde{X}}(u) = L\big(u, \operatorname{Ind}_H^G 1\big) = \prod_{\kappa \in \widehat{G}} L(u, \kappa)^{m_\kappa}.$$

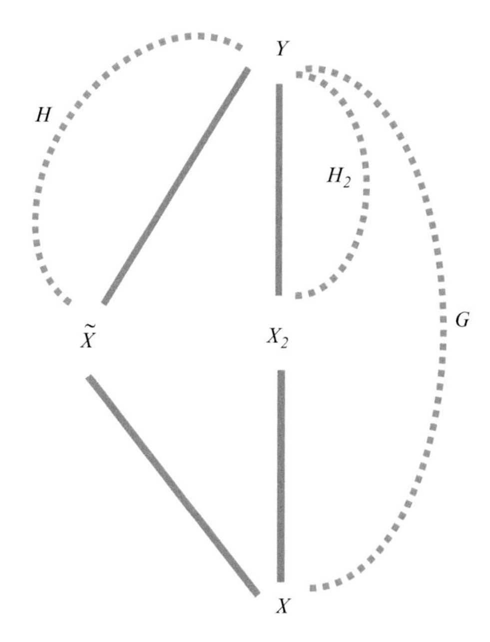

Figure 23.2 The covering appearing in Theorem 23.5. The Galois groups are shown by the dashed lines.

The Artin–Ihara L-function $L(u,\kappa)$ appears m_κ times in the factorization, and Frobenius reciprocity (see Theorem 18.4) says that

$$m_\kappa = \left\langle \chi_{\mathrm{Ind}_H^G 1}, \kappa \right\rangle = \langle 1, \kappa|_H \rangle \leq \deg \kappa.$$

Let $\kappa = \pi_2$, for which $\deg\kappa = 1$. This implies that $\zeta_{\widetilde{X}}(u)$ has $-R_X$ as a (simple) pole if and only if $\pi_2|_H$ is the identity. Note that, for $\pi \neq \kappa$, $-R_X$ is not a pole of any $L(u,\pi)$. We have $\pi_2|_H$ is the identity if and only if $H \subset H_2 = \ker\pi_2$, which is equivalent to saying that $\widetilde{X}$ covers X_2. Here we use part (5) of the fundamental theorem of Galois theory, Theorem 14.3.

Finally, X_2 is unique as each version of X_2 would cover the other. $\square$

Note that, in Theorem 23.5, if $\widetilde{X}$ is intermediate to Y/X then $\Delta(\widetilde{X}) = 1$ or 2 and the Perron–Frobenius theorem says that $\zeta_{\widetilde{X}}(u)$ is a function of u^d, where d is the number of poles of $\zeta_{\widetilde{X}}(u)$ on the circle $|u| = R$. Thus the $\widetilde{X}$ for which $\Delta(\widetilde{X}) = 2$ are exactly those for which $\zeta_{\widetilde{X}}(u)$ has $-R$ as a Siegel pole, which are the $\widetilde{X}$ that cover X_2. Since $\Delta(\widetilde{X}) = 2$ is the condition for $\widetilde{X}$ to be bipartite, this says that $\widetilde{X}$ is bipartite. The intermediate graph $\widetilde{X}$ is not quadratic unless $\widetilde{X} = X_2$. All other graphs $\widetilde{X}$ intermediate to Y/X have $\Delta_{\widetilde{X}} = 1$.

Every graph X satisfying our usual hypotheses has a covering Y whose zeta function has a Siegel pole (**exercise**). This is probably not the case for algebraic number fields.

Corollary 23.8 *Under the hypotheses of Theorem 23.5, with X_2 the unique graph defined in that theorem, the set of intermediate bipartite covers of Y/X is precisely the set of graphs intermediate to Y/X_2.*

23.3 General case; inflation and deflation

To understand the case $\delta > 1$ we need some definitions.

Definition 23.9 An **inflation** graph $I^{\delta}(X)$ is defined as the graph resulting from adding $\delta - 1$ vertices to every edge of X.

Definition 23.10 The **deflation** graph $D_{\delta}(X)$ is obtained from X by collapsing δ consecutive edges between consecutive nodes to one edge.

See Figure 23.3 for an example.

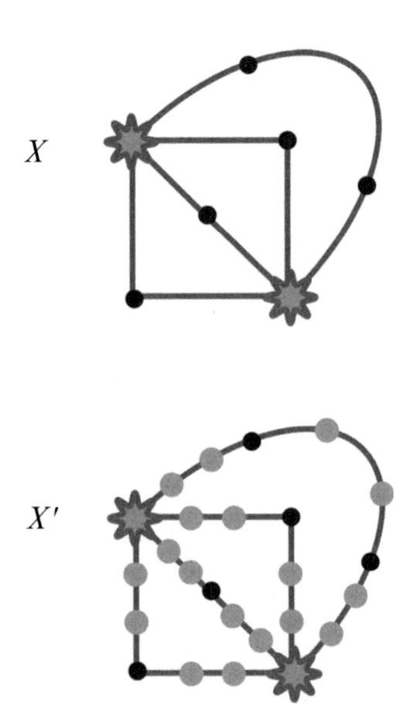

Figure 23.3 The graph X' is an inflation of X increasing the length of paths by a factor 3, while the graph X is a deflation of X'.

Theorem 23.11 (Siegel poles in general) *Suppose that X and a cover Y of X satisfy our usual hypotheses and $\delta = \delta_X = \Delta_X$. Suppose that $\Delta_Y = 2\Delta_X = 2\delta$. Then we have the following facts.*

(1) There is a unique quadratic cover X_2 intermediate to Y/X such that $\Delta_{X_2} = 2\delta$.
(2) Let $\widetilde{X}$ be any graph intermediate to Y/X. Then $\Delta_{\widetilde{X}} = 2\delta$ if and only if $\widetilde{X}$ is intermediate to Y/X_2.

Proof When $\delta > 1$ this is proved by deflation. The deflated graph $D_\delta(X) = X'$ contains all the information on X and its covers. This graph X' has $\delta_{X'} = 1$ and $\zeta_X(u) = \zeta_{X'}(u^\delta)$. Every single Y/X has a corresponding deflated covering Y' of X' such that

$$\zeta_Y(u) = \zeta_{Y'}(u^{\delta_X}).$$

There is also a relation between all the Artin L-functions:

$$L_{Y/X}(u, \pi) = L_{Y'/X'}(u^{\delta_X}, \pi).$$

Here π is a representation of $\mathrm{Gal}(Y/X) = \mathrm{Gal}(Y'/X')$. Theorem 23.11 now follows from Theorem 23.5. □

If $\delta = 1$ in Theorem 23.11, the graphs $\widetilde{X}$ with $\Delta_{\widetilde{X}} = 2\delta$ are the bipartite covering graphs intermediate to Y/X and, in particular, X_2 is bipartite. Even when $\Delta_X = \delta_X = \delta$ is odd, the graphs $\widetilde{X}$ with $\Delta_{\widetilde{X}} = 2\delta$ are precisely the bipartite covering graphs intermediate to Y/X. But if $\Delta_X = \delta_X = \delta$ is even then every graph intermediate to Y/X, including X itself, is bipartite and thus, being bipartite, does not determine which quadratic cover of X is X_2. Note also that when the rank of X is ≥ 2, we have proved the following purely graph theoretic equivalent theorem. For a combinatorial proof of the following theorem, see [122].

Theorem 23.12 (The story of bipartite covers) *Suppose that X is a finite connected graph of rank ≥ 1 and that Y is a bipartite covering graph of X. Then we have the following facts.*

(1) When X is bipartite, every intermediate covering $\widetilde{X}$ to Y/X is bipartite.
(2) When X is not bipartite, there is a unique quadratic covering graph X_2 intermediate to Y/X such that any graph $\widetilde{X}$ intermediate to Y/X is bipartite if and only if $\widetilde{X}$ is intermediate to Y/X_2.

Example 23.13 Let $X = K_4$ and Y be the cube. Then:

(1) $\Delta_X = 1 = \delta_X, \delta_Y = 1, \Delta_Y = 2$;
(2) neither ζ_X nor ζ_Y has a Siegel pole;
(3) $X_2 = Y = Y_2$.

Exercise 23.3 Perform the same calculations as in the last example for $Y = Y_6$, the S_3 cover of $X = K_4 - e$ in Figure 14.4.

Part V

Last look at the garden

In Part V we examine applications of zeta functions to error-correcting codes, explicit formulas analogous to Selberg's trace formula, further connections with random matrix theory, and finally some research problems.

24

An application to error-correcting codes

Unlike cryptographic codes, error-correcting codes are used to make a message understandable even though it has been corrupted by some problem in transmission or recording. References for the subject include Vera Pless [101] and Audrey Terras [133], Chapter 11. The application of edge zetas that we are considering comes from the papers of Ralf Koetter, Winnie Li, Pascal Vontobel, and Judy Walker [70], [71].

Definition 24.1 A binary $[n, k]$ linear **code** C is a k-dimensional subspace of the vector space $\mathbb{F}_2^n$, where $\mathbb{F}_2$ denotes the field with two elements.

One way to specify such a code C uses the **parity check** or Hamming matrix H:

$$C = \left\{x \in \mathbb{F}_2^n \mid Hx = 0\right\}.$$

If H is an $s \times n$ matrix, it must have rank $n - k$, in order for C to be an $[n, k]$ code.

We will consider an example from [70], [71].

Example 24.2 A **parity check matrix** for a $[7, 2]$ code is the matrix H given by

$$H = \begin{pmatrix} 1 & 1 & 0 & 0 & 0 & 0 & 0 \\ 0 & 1 & 1 & 1 & 0 & 0 & 0 \\ 1 & 0 & 1 & 0 & 0 & 0 & 0 \\ 0 & 0 & 0 & 1 & 1 & 0 & 1 \\ 0 & 0 & 0 & 0 & 1 & 1 & 0 \\ 0 & 0 & 0 & 0 & 0 & 1 & 1 \end{pmatrix}.$$

Next we define the **Tanner graph** $T(H)$ associated with the above parity check matrix H. Assume that H is an $s \times n$ matrix. The graph $T(H)$ is a bipartite graph with **bit** vertices $X_1, \ldots, X_n$ and **check** vertices $p_1, \ldots, p_s$. Connect p_i to X_j with an edge iff the parity check matrix has ij entry $h_{ij} = 1$. The Tanner graph for the example above is shown in Figure 24.1 along with a quadratic covering.

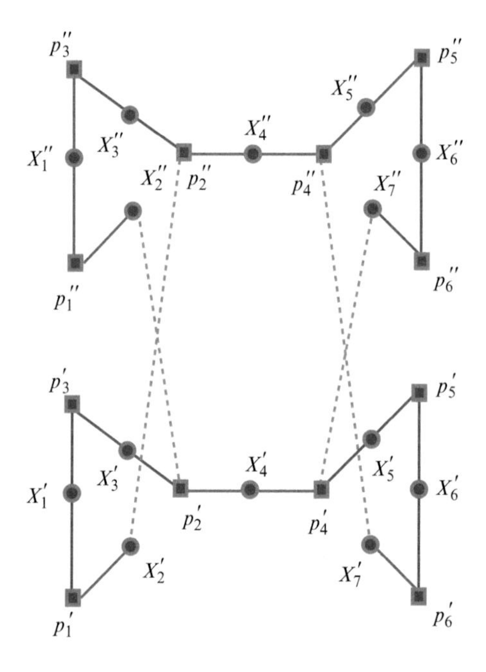

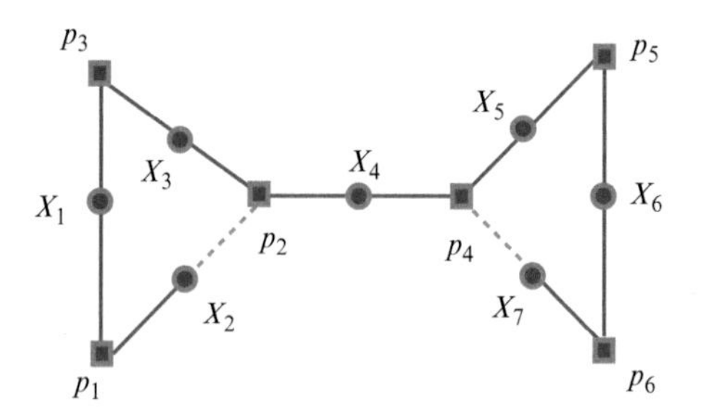

Figure 24.1 The Tanner graph (*lower diagram*) of the code corresponding to the parity check matrix H in Example 24.2 is shown along with a quadratic cover (*upper diagram*). The sheets of the cover are shown by solid lines. The edges left out of the spanning tree for $T(H)$ are shown by dashed lines, as are their lifts to the cover.

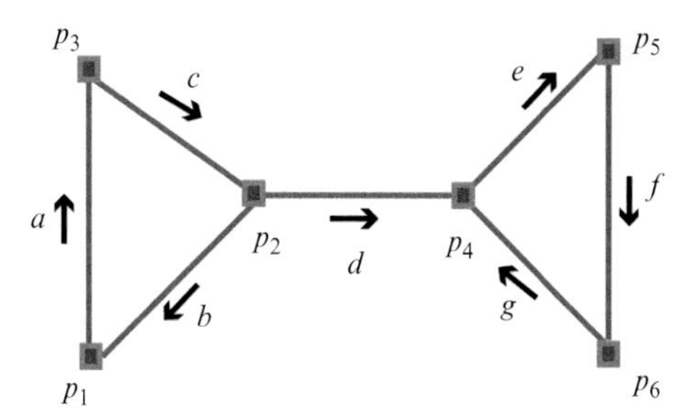

Figure 24.2 The normal graph $N(H)$ corresponding to the Tanner graph $T(H)$ in the lower diagram in Figure 24.1.

We call the code corresponding to the parity check matrix H in Example 24.2 a **cycle code** since each bit vertex X_j in $T(H)$ in Figure 24.1 has degree 2. The **normal graph** $N(H)$ associated with H is obtained by collapsing or deflating the edges, as in Definition 23.10, with the bit nodes X_j in $T(H)$ to just an edge. Figure 24.2 shows the normal graph $N(H)$ for the parity check matrix in Example 24.2.

Of course, the usefulness of error-correcting codes has much to do with the efficiency of the decoding algorithms. Iterative decoding operates locally and thus has trouble with codewords from coverings. This leads to the definition of a pseudo-codeword given below. One of the main results of Koetter *et al.* [70], [71] says that edge zetas give a list of pseudo-codewords.

Write the bit vertices of the quadratic cover in the order $X_1', X_1'', \ldots, X_n', X_n''$. List the check vertices in a similar manner. The quadratic covering in Figure 24.1 corresponds to a code whose parity check matrix is given by the matrix with block form

$$\widetilde{H} = \begin{pmatrix} I & I & 0 & 0 & 0 & 0 & 0 \\ 0 & J & I & I & 0 & 0 & 0 \\ I & 0 & I & 0 & 0 & 0 & 0 \\ 0 & 0 & 0 & I & I & 0 & J \\ 0 & 0 & 0 & 0 & I & I & 0 \\ 0 & 0 & 0 & 0 & 0 & I & I \end{pmatrix}.$$

Here

$$I = \begin{pmatrix} 1 & 0 \\ 0 & 1 \end{pmatrix}, \qquad J = \begin{pmatrix} 0 & 1 \\ 1 & 0 \end{pmatrix}, \qquad 0 = \begin{pmatrix} 0 & 0 \\ 0 & 0 \end{pmatrix}.$$

Every element c of the original code C lifts to an element $\tilde{c}$ of the covering code $\widetilde{C}$. For example, the code word (1110000) in our example code lifts to

(11 11 11 00 00 00 00). However, the codeword $\tilde{c} = (10\ 10\ 10\ 11\ 10\ 10\ 10)$ in $\widetilde{C}$ is not a lift of a codeword in C.

Exercise 24.1 Using the preceding definitions of H and $\widetilde{H}$, check that $H^t(1110000) = 0$. and $\widetilde{H}^t(10\ 10\ 10\ 11\ 10\ 10\ 10) = 0$.

Definition 24.3 Suppose that our original $[n, k]$ code is C, with corresponding Tanner graph $T(H)$. The **unscaled pseudo-codeword** corresponding to a codeword $\tilde{c}$ coming from an M-sheeted covering of the Tanner graph $T(H)$ is defined to be $\omega(\tilde{c}) \in \mathbb{Z}^n$, with jth entry obtained by summing the M entries above c_j in $\tilde{c}$. Note that we do not sum in the finite field but instead sum the 0's and 1's as ordinary integers. The **normalized pseudo-codeword** is obtained by dividing $\omega(\tilde{c})$ by M. It lies in $\mathbb{Q}^n$.

Thus, for example, consider the codeword $\tilde{c} = (10\ 10\ 10\ 11\ 10\ 10\ 10)$ coming from the quadratic cover in Figure 24.1. Then the corresponding pseudo-codeword is $\omega(\tilde{c}) = (1\ 1\ 1\ 2\ 1\ 1\ 1)$. The normalized version is $\left(\frac{1}{2}, \frac{1}{2}, \frac{1}{2}, 1, \frac{1}{2}, \frac{1}{2}, \frac{1}{2}\right)$. The normalized pseudo-codeword coming from the lift of a codeword in C will consist entirely of 0's and 1's, and conversely.

The key problem with iterative decoding is that the algorithm cannot distinguish between codewords coming from finite covers of $T(H)$. Thus one becomes interested in pseudo-codewords. See the website

http://www.hpl.hp.com/personal/Pascal_Vontobel/pseudocodewords/papers/papers_pseudocodewords.html

for more information on pseudocodewords.

The following theorem is proved in [70] for cycle codes and in [71] more generally. We will restrict our considerations to cycle codes here.

Theorem 24.4 (Ralf Koetter, Winnie Li, Pascal Vontobel, and Judy Walker) *Let C be an $[n, k]$ cycle code with parity check matrix H and normal graph $N = N(H)$. Let $\zeta_N(u_1, \ldots, u_n)$ denote the edge zeta of N with edge matrix W specialized to have ef entry $W_{ef} = u_i$, where e is a directed edge corresponding to the ith edge of N (i.e., the edge corresponding to the ith bit vertex X_i). Then the monomial $u_1^{a_1} \cdots u_n^{a_n}$ appears with a non-zero coefficient in $\zeta_N(u_1, \ldots, u_n)$ iff the corresponding exponent vector $(a_1, \ldots, a_n)$ is an unscaled pseudo-codeword for C.*

Let us consider our example again. The zeta function of the normal graph $N(H)$ is found by computing

$$
\zeta_N(a,b,c,d,e,f,g)^{-1}
$$

$$
= \det \begin{pmatrix}
-1 & 0 & a & 0 & 0 & 0 & 0 & 0 & 0 & 0 & 0 & 0 & 0 & 0 \\
b & -1 & 0 & 0 & 0 & 0 & 0 & 0 & 0 & 0 & 0 & 0 & 0 & 0 \\
0 & c & -1 & c & 0 & 0 & 0 & 0 & 0 & 0 & 0 & 0 & 0 & 0 \\
0 & 0 & 0 & -1 & d & 0 & 0 & 0 & 0 & 0 & 0 & 0 & 0 & d \\
0 & 0 & 0 & 0 & -1 & e & 0 & 0 & 0 & 0 & 0 & 0 & 0 & 0 \\
0 & 0 & 0 & 0 & 0 & -1 & f & 0 & 0 & 0 & 0 & 0 & 0 & 0 \\
0 & 0 & 0 & 0 & g & 0 & -1 & 0 & 0 & 0 & g & 0 & 0 & 0 \\
0 & 0 & 0 & 0 & 0 & 0 & 0 & -1 & a & 0 & 0 & 0 & 0 & 0 \\
0 & 0 & 0 & b & 0 & 0 & 0 & 0 & -1 & b & 0 & 0 & 0 & 0 \\
0 & 0 & 0 & 0 & 0 & 0 & 0 & c & 0 & -1 & 0 & 0 & 0 & 0 \\
0 & d & 0 & 0 & 0 & 0 & 0 & 0 & 0 & d & -1 & 0 & 0 & 0 \\
0 & 0 & 0 & 0 & 0 & 0 & 0 & 0 & 0 & 0 & e & -1 & 0 & e \\
0 & 0 & 0 & 0 & 0 & 0 & 0 & 0 & 0 & 0 & 0 & f & -1 & 0 \\
0 & 0 & 0 & 0 & 0 & 0 & 0 & 0 & 0 & 0 & 0 & 0 & g & -1
\end{pmatrix}
$$

$$
\begin{aligned}
&= -4a^2b^2c^2d^2e^2f^2g^2 + 4a^2b^2c^2d^2efg + a^2b^2c^2e^2f^2g^2 - 2a^2b^2c^2efg \\
&\quad + 4abcd^2e^2f^2g^2 - 4abcd^2efg - 2abce^2f^2g^2 + 4abcefg \\
&\quad + e^2f^2g^2 - 2efg + a^2b^2c^2 - 2abc + 1.
\end{aligned}
$$

The first few terms in the power series for $\zeta_N(a,b,c,d,e,f,g)$ are

$$
\begin{aligned}
&1 + 2abc + 2efg + 3a^2b^2c^2 + 3e^2f^2g^2 + 4abcefg + 4abcd^2efg \\
&\quad + 6abce^2f^2g^2 + 6a^2b^2c^2efg + 12abcd^2e^2f^2g^2 + 12a^2b^2c^2d^2efg \\
&\quad + 9a^2b^2c^2e^2f^2g^2 + 36a^2b^2c^2d^2e^2f^2g^2.
\end{aligned}
$$

The degrees of the terms are 0, 3, 3, 6, 6, 6, 8, 9, 9, 11, 11, 12, 14. According to the preceding theorem, these terms lead to the pseudo-codewords

$$
\begin{array}{lll}
(0,0,0,0,0,0,0), & (1,1,1,0,0,0,0), & (0,0,0,0,1,1,1), \\
(2,2,2,0,0,0,0), & (0,0,0,0,2,2,2), & (1,1,1,0,1,1,1), \\
(1,1,1,2,1,1,1), & (1,1,1,0,2,2,2), & (2,2,2,0,1,1,1), \\
(1,1,1,2,2,2,2), & (2,2,2,2,1,1,1), & (2,2,2,0,2,2,2), \\
(2,2,2,2,2,2,2). & &
\end{array}
$$

25
Explicit formulas

We can also produce analogs of the explicit formulas of analytic number theory. That is, we seek an analog of Weil's explicit formula for the Riemann zeta function. In his original work Weil used that result to formulate a statement equivalent to the Riemann hypothesis. See Weil [141]. One can also view such explicit formulas as analogs of Selberg's trace formula, the result used to study Selberg's zeta function, which we defined in formula (3.1).

Our analog of the Von Mangoldt function from elementary number theory is N_m. Using formula (4.4), we have

$$u\frac{d}{du}\log\zeta(u,X) = -u\frac{d}{du}\sum_{\lambda\in \mathrm{Spec}\, W_1}\log(1-\lambda u)$$

$$= \sum_{\lambda\in \mathrm{Spec}\, W_1}\frac{\lambda u}{1-\lambda u} = -\sum_{\rho \text{ a pole of } \zeta}\frac{u}{u-\rho}. \tag{25.1}$$

Then it is not hard to prove the following result by following the method of Murty [91], p. 109.

Proposition 25.1 (An explicit formula) *Let $0 < a < R$, where R is the radius of convergence of $\zeta(u, X)$. Assume that $h(u)$ is meromorphic in the plane and holomorphic outside the circle of center 0 and radius $a - \varepsilon$, for small $\varepsilon > 0$. Assume also that $h(u) = O(|u|^p)$ as $|u| \to \infty$ for some $p < -1$. Finally, assume that the transform $\widehat{h}_a(n)$ of $h(u)$ decays rapidly enough for the right-hand side of the formula to converge absolutely. Then, if N_m is as in Definition 4.2, we have*

$$\sum_{\rho}\rho h(\rho) = \sum_{n\geq 1} N_n \widehat{h}_a(n),$$

where the sum on the left is over the poles of $\zeta(u, X)$ and

$$\widehat{h}_a(n) = \frac{1}{2\pi i} \oint_{|u|=a} u^n h(u) du.$$

Proof We follow the method of Murty [91], p. 109. Look at

$$\frac{1}{2\pi i} \oint_{|u|=a} \left\{ u \frac{d}{du} \log \zeta(u, X) \right\} h(u) du.$$

Use Cauchy's integral formula to move the contour over to the circle $|u| = b > 1$. Then let $b \to \infty$. Also use formulas (25.1) and (10.1). Note that $N_n \sim \Delta_X / R_X^m$ as $m \to \infty$. □

Such explicit formulas are basic for work on the pair correlation of complex zeros of zeta (see Montgomery [90]). They can also be viewed as an analog of Selberg's trace formula. See [57] and [136] for discussions of Selberg's trace formula for a $(q+1)$-regular graph. In these papers various kernels (e.g., Green's, the characteristic functions of intervals, heat) were plugged into the trace formula and various results such as McKay's theorem on the distribution of eigenvalues of the adjacency matrix and the Ihara determinant formula for the Ihara zeta were deduced. It would be an interesting research project to do the same sort of investigation for irregular graphs.

Exercise 25.1 Plug the function $h(u) = u^{-m-1}$, $m = 1, 2, 3, \ldots$ into the explicit formula. You should get a result that is well known to us.

26
Again chaos

In our earlier chapter on chaos, Chapter 5, we considered the spacing of zeta poles of regular graphs as well as the eigenvalue distribution for the adjacency matrices of regular graphs. See Figure 5.6. In particular, we saw in Table 5.1 that there is a conjectural dichotomy between the behavior of the zetas of random regular graphs and that of the zetas of Cayley graphs of abelian groups, for example.

In this chapter we investigate the spacings of the poles of the Ihara zeta function of a random irregular graph and compare the result with the spacings for covering graphs, both random and also with abelian Galois group. By formula (4.4) this is essentially the same as investigating the spacings of the eigenvalues of the edge adjacency matrix W_1 from Definition 4.1. Here, although W_1 is not symmetric, the nearest neighbor spacing can be studied. If the eigenvalues of the matrix are λ_i, $i = 1, \ldots, 2m$, we want to look at

$$v_i = \min\left\{|\lambda_i - \lambda_j| \,\middle|\, j \neq i\right\}.$$

The question becomes: which function best approximates the histogram of the v_i, assuming that they are normalized to have mean 1?[1]

References for the study of the spacings of the eigenvalues of non-symmetric matrices include Ginibre [43], LeBoeuf [74], and Mehta [85]. The **Wigner surmise for non-symmetric matrices** is that this spacing is given by

$$4\Gamma\left(\frac{5}{4}\right)^4 x^3 \exp\left(-\Gamma\left(\frac{5}{4}\right)^4 x^4\right). \tag{26.1}$$

[1] In the figures in this chapter all spacings have been normalized to have mean 1.

Before looking at spacings coming from the poles of a zeta function, we will consider the eigenvalues of a random matrix with block form $\begin{pmatrix} A & B \\ C & {}^{t}A \end{pmatrix}$ where B and C are symmetric matrices that have zeros down the diagonal. This is the type of matrix that we get for the edge adjacency matrix W_1 of a graph, as we saw in Chapter 11. We can use Matlab's `randn(N)` command to get matrices A, B, C with normally distributed entries. A result known as the Girko circle law says that the eigenvalues of a set of random $n \times n$ real matrices with independent entries having a standard normal distribution should be approximately uniformly distributed in a circle of radius $\sqrt{n}$ for large n. References are Bai [6], Girko [44], and Tao and Vu [131]. A plot of the eigenvalues of a random matrix having the properties of W_1 is to be found in Figure 26.1. Note the symmetry with respect to the real axis, owing

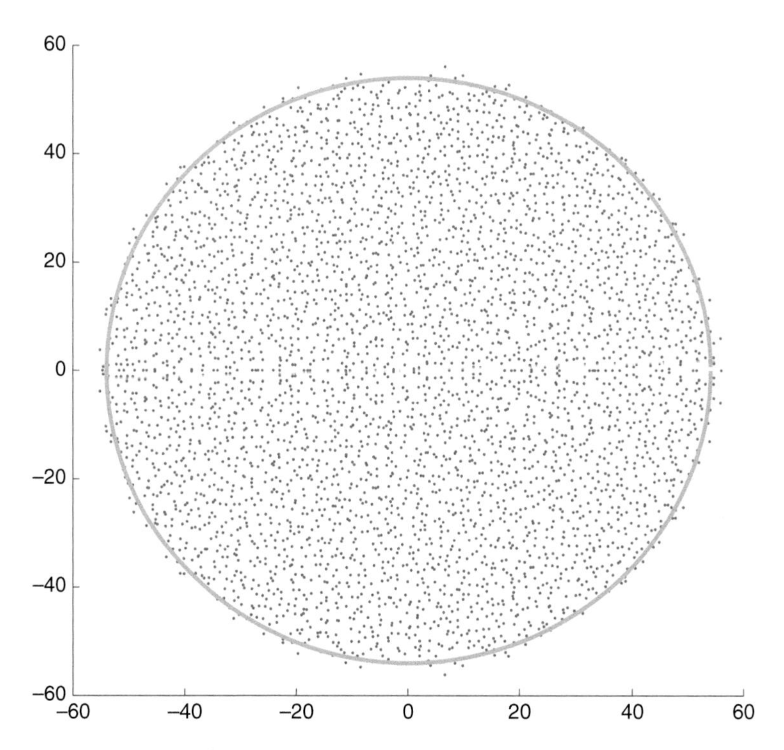

Figure 26.1 The results of a Matlab experiment, showing the spectrum of a random 2000×2000 matrix with the properties of W_1 except that the entries are not 0 and 1. The circle has radius $r = \frac{1}{2}(1 + \sqrt{2})\sqrt{2000}$, rather than $\sqrt{2000}$ as in Girko's circle law.

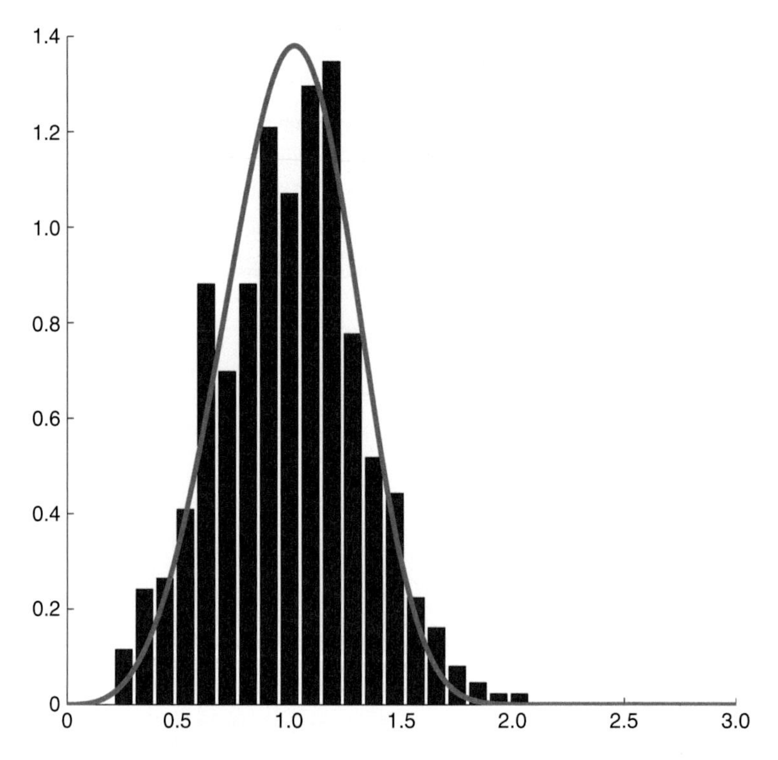

Figure 26.2 The histogram shows the normalized nearest neighbor spacing for the spectrum given in Figure 26.1. The curve is the Wigner surmise from formula (26.1).

to the fact that our matrix is real. Another interesting fact is that the circle radius is not exactly that which Girko predicts. The spacing distribution for this random matrix is seen to be close to the non-symmetric Wigner surmise (26.1) in Figure 26.2.

Figure 11.2 in the chapter on the edge zeta shows the results of a Matlab experiment giving the spectrum of the edge adjacency matrix W_1 for a "random graph." Figure 26.3 shows a histogram of the nearest neighbor spacings of the spectrum of the random graph from Figure 11.2 along with two curves showing the **modified Wigner surmise**

$$(\omega+1)\Gamma\left(\frac{\omega+2}{\omega+1}\right)^{\omega+1} x^{\omega}\exp\left(-\Gamma\left(\frac{\omega+2}{\omega+1}\right)^{\omega+1} x^{\omega+1}\right) \qquad (26.2)$$

for $\omega = 3$ and 6. When $\omega = 3$, (26.2) is the original Wigner surmise (26.1).

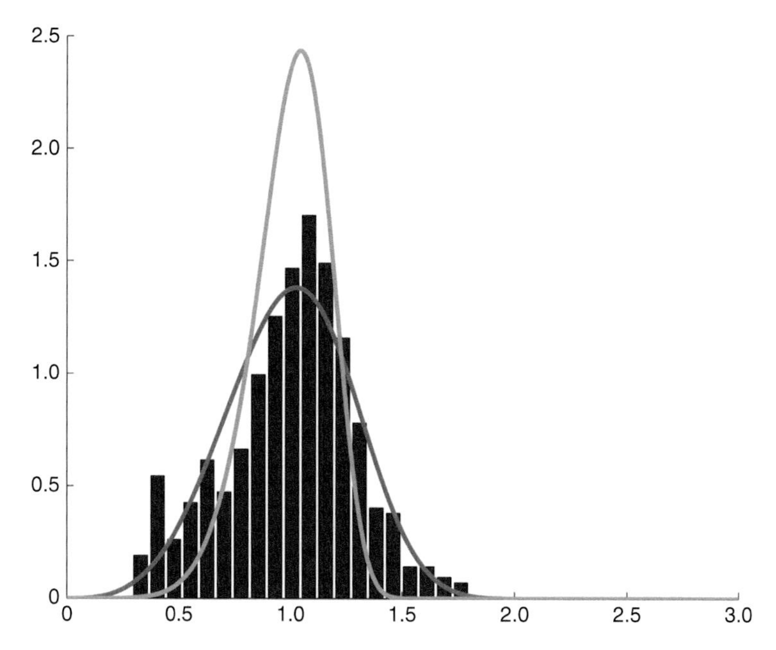

Figure 26.3 The histogram shows the nearest neighbor spacings of the spectrum of the random graph from Figure 11.2. The curves give the modified Wigner surmise from formula (26.2) for $\omega = 3$ and 6.

For covering graphs, one can say more about the expected shape of the spectrum of the edge adjacency matrix or, equivalently, describe the region bounding the poles of the Ihara zeta. Angel, Friedman, and Hoory [2] give a method for computing the region encompassing the spectrum of the operator analogous to the edge adjacency matrix W_1 on the universal cover of a graph X. In Chapter 2, we mentioned the Alon conjecture for regular graphs. Angel, Friedman, and Hoory give an **analog of the Alon conjecture for irregular graphs**. Roughly, their conjecture says that the new edge adjacency spectrum of a large random covering graph is near the edge adjacency spectrum of the universal covering. Here "new" means "not occurring in the spectrum of W_1 for the base graph." This conjecture can be shown to imply the approximate Riemann hypothesis for the new poles of a large random cover.

The truth of the Angel, Friedman, and Hoory analog of the Alon conjecture is visible from Figure 26.4, drawn by Tom Petrillo [100] using the methods of Angel, Friedman, and Hoory. This figure shows a large random cover of the base graph X consisting of two loops with an extra vertex on one loop. The light blue region shows the spectrum of the edge adjacency operator on the universal cover of the base graph X.

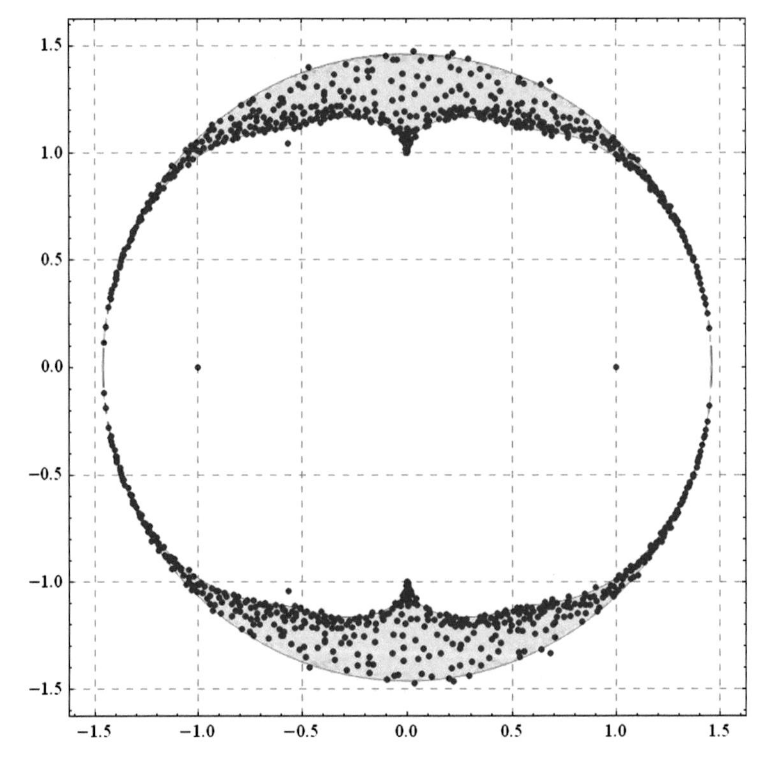

Figure 26.4 Tom Petrillo's figure (from [100], produced using Mathematica) showing in light blue the region bounding the spectrum of the edge adjacency operator on the universal cover of the base graph consisting of two loops with an extra vertex on one loop. The darker blue points are the eigenvalues of the edge adjacency matrix of a large random graph. The Angel, Friedman, and Hoory analog of the Alon conjecture for irregular graphs appears to be valid.

We will give some other examples related to this analog of the Alon conjecture. Figure 26.6 shows the spectrum of the edge adjacency matrix of a random cover of the base graph consisting of two loops with an extra vertex on one loop. The smallest circle has radius 1, the middle circle has radius $1/\sqrt{R}$, and the largest circle has radius $\sqrt{3}$. The Riemann hypothesis is approximately true for this graph zeta as is the analog of the Alon conjecture made by Angel, Friedman, and Hoory.

To produce a figure such as Figure 26.6 for a random cover Y of the base graph X consisting of two loops with an extra vertex on one of them, we can use the formula for the edge adjacency matrix W_1 of Y in terms of the start matrix S and the terminal matrix T from Proposition 11.7. It is also convenient to write $S = (MN)$, $T = (NM)$, where M and N each have $m = |E|$ columns. We used this fact in the proof in Chapter 11 of the Kotani and Sunada theorem, Theorem 8.1. It follows that

$$W_1 = \begin{pmatrix} {}^{t}NM & {}^{t}NN - I \\ {}^{t}MM - I & {}^{t}MN \end{pmatrix}.$$

Now we arrange the columns of M so that columns corresponding to lifts of a given edge of the base graph are listed in order of the sheet on which the lift starts. The lifts of a given vertex of the base graph are also listed together in the order of the sheets where they live. Then

$$M = \begin{pmatrix} I_n & I_n & 0 \\ 0 & 0 & I_n \end{pmatrix} \quad \text{and} \quad N = \begin{pmatrix} A & 0 & I_n \\ 0 & B & 0 \end{pmatrix},$$

where A and B are permutation matrices. Suppose that $n = 3$ and the lift of edge a corresponds to the permutation (12) while the lift of edge b corresponds to the permutation (13). Then we get the graph in Figure 26.5.

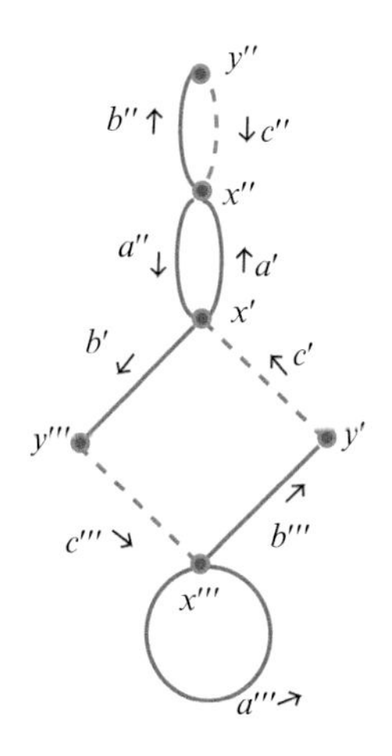

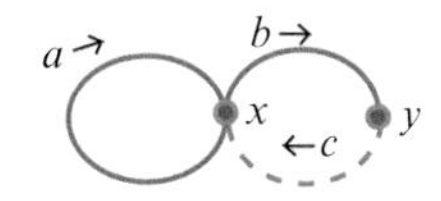

Figure 26.5 The random 3-cover of two loops with an extra vertex on one loop; the lift of a corresponds to the permutation (12) and the lift of b corresponds to the permutation (13). The dashed line in the base graph is the spanning tree and the dashed lines in the cover are the sheets of the cover.

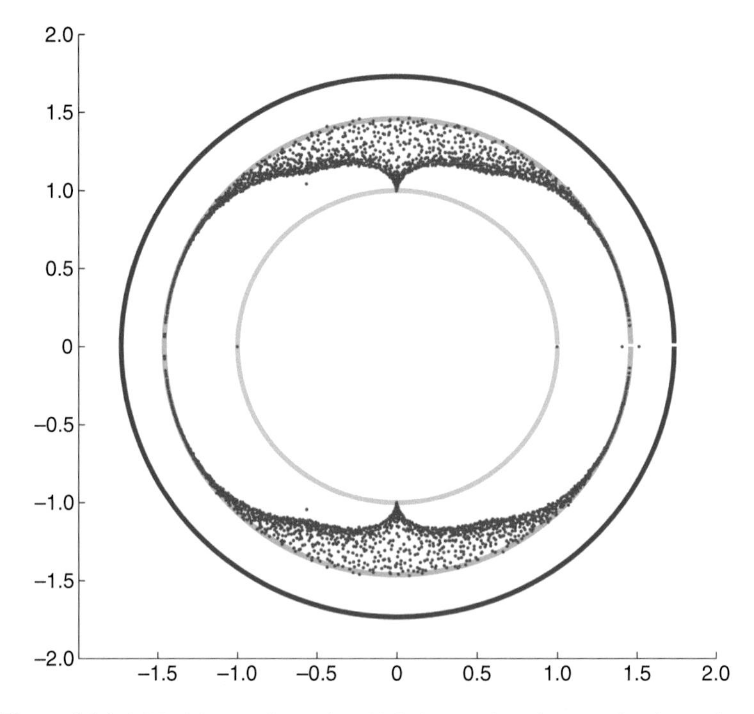

Figure 26.6 A Matlab experiment in which the purple points are the eigenvalues of the edge adjacency matrix of a random cover (with 801 sheets) of the base graph consisting of two loops with an extra vertex on one loop. Thus the plot shows the reciprocals of the poles of the zeta function of the covering. The innermost circle has radius 1, the middle circle has radius $1/\sqrt{R}$, and the outermost circle has radius $\sqrt{3}$. The Riemann hypothesis is approximately true.

We used Matlab to plot the eigenvalues of W_1 for covers in which A and B are random permutation matrices (found using the command `randperm` in Matlab). With $n = 801$, we obtain the spectrum in Figure 26.6. If this is compared with the picture found by Angel, Friedman, and Hoory [2] for random covers of the base graph $K_4 - e$, where e is an edge, we can see that there is much similarity even though their Frobenius eigenvalue is 1.5 while ours is approximately 2.1304. It should also be compared with Figure 26.4.

Figure 26.7 shows the nearest neighbor spacings for the points in Figure 26.6 compared with the modified Wigner surmise in formula (26.2), for various small values of ω.

Figure 26.8 shows the spectrum of the edge adjacency matrix for a Galois $\mathbb{Z}_{163} \times \mathbb{Z}_{45}$ covering of a base graph consisting of two loops with an extra

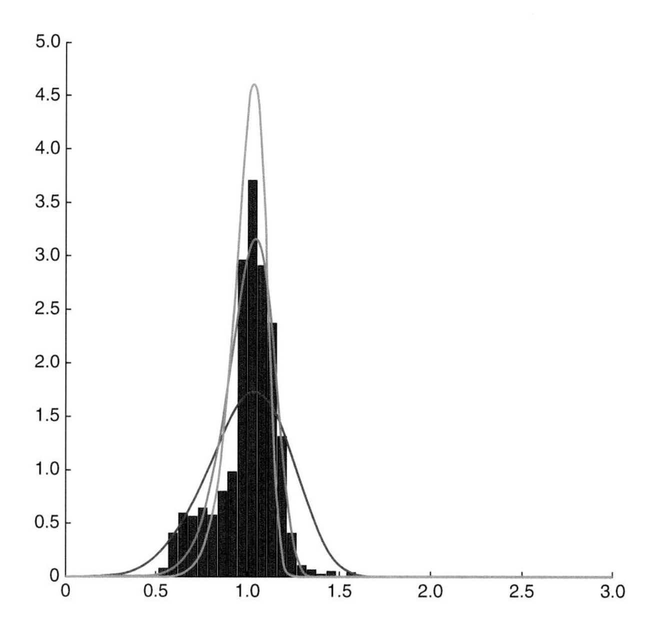

Figure 26.7 A histogram of the nearest neighbor spacings for the spectrum of the edge adjacency matrix of Figure 26.6. The curves give three versions of the modified Wigner surmise (26.2). Here $\omega = 3, 6,$ and 9.

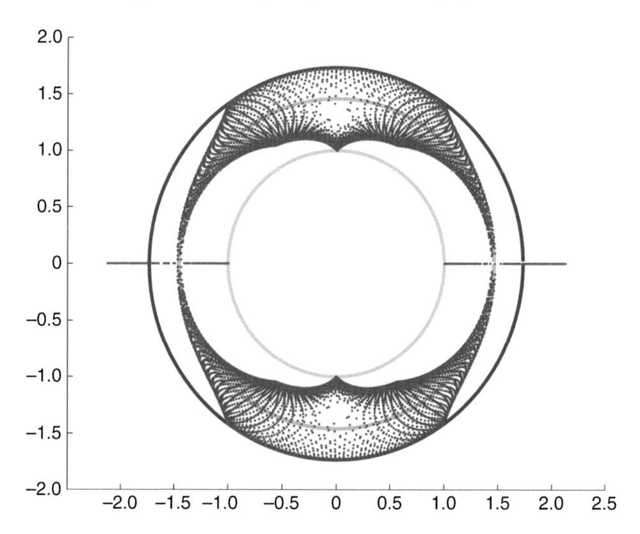

Figure 26.8 The results of a Matlab experiment in which the purple dots are the eigenvalues of the edge adjacency matrix W_1 for a Galois $\mathbb{Z}_{163} \times \mathbb{Z}_{45}$ covering of the graph consisting of two loops with an extra vertex on one loop. The innermost circle has radius 1, the middle circle has radius $1/\sqrt{R}$, and the outermost circle has radius $\sqrt{3}$. The Riemann hypothesis is very false.

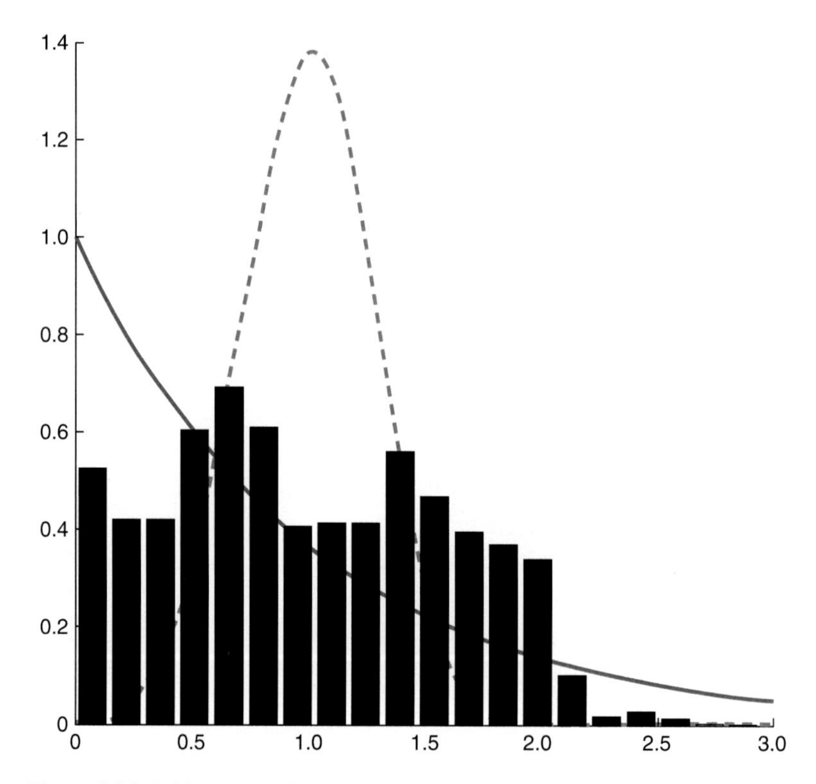

Figure 26.9 A histogram of the nearest neighbor spacings for the spectrum of the edge adjacency matrix W_1 of the graph in Figure 26.8. The spacing distribution e^{-x} for a Poisson random variable and a dashed curve showing the Wigner surmise (26.1) are also shown.

vertex on one loop. The innermost circle has radius 1, the middle circle has radius $1/\sqrt{R}$ with R as in Definition 2.4. The outermost circle has radius $\sqrt{3}$. The Riemann hypothesis looks very false.

Figure 26.9 shows a histogram of the nearest neighbor spacings for the spectrum of the edge adjacency matrix of the graph in Figure 26.8 with, for comparison, the spacing distribution for a Poisson random variable (e^{-x}) and the Wigner surmise (26.1). This should be compared with the spacings of the Laplacian eigenvalues for $\Gamma\backslash H$, where Γ is the arithmetic group and H is the upper half plane.

27

Final research problems

I leave the reader with my list of possible research projects.

(1) Do experiments on the differences between the properties of zetas of weighted or quantum graphs and unweighted graphs. See Horton, Stark, and Terras [60], [61]. In particular, consider the connections with random matrix theory. See also Smilansky [115].

(2) (a) Investigate the poles of the Ihara zeta and the Riemann hypothesis for random graphs. How does the distribution of poles depend on the probability of an edge and on the degree sequence?
(b) Can one find a Galois graph covering Y of a base graph X such that the poles of the Ihara zeta of Y behave like those of a random cover of X? One could experiment with various finite non-abelian groups. We considered abelian groups in the preceding chapters; the pole distributions of zetas for abelian covers were seen to be very different from those of random covers. One idea is to imitate the work of Lubotzky, Phillips, and Sarnak [79] using the group $\mathrm{SL}(2, F)$, where F is a finite field.

(3) Can you prove an analog of the theorem of Katz and Sarnak [68] for function field zeta functions at least in the case of regular graphs? This theorem says that, for almost all curves over a finite field, as the genus and order of the field go to infinity, the imaginary parts of the zeros of zetas approach the GUE level spacing (meaning that the spacings look like those Odlyzko found for the high zeros of the Riemann zeta, as in Figure 5.5. An elementary reference giving background on this subject is the book by Miller and Takloo-Bighash [86]. See their bibliography on the web too! Examples and easier proofs should be found for a graph theory version. One expects pole spacings of regular graph zetas to be related to GOE spacings (i.e., those for the eigenvalues of real symmetric matrices). See the experiments of D. Newland, shown in Figure 5.6

and discussed in Chapter 5. Equivalently, one expects the level spacings for the eigenvalues of the adjacency matrices in a sequence of regular graphs satisfying the hypotheses of McKay's theorem 5.3 to approach those for the GOE as $n \to \infty$.

(4) Figure out what a ramified covering is and how the zeta function of such a covering factors. More on this question is found in Section 13.3 on coverings and in Malmskog and Manes [81] or Baker and Norine [8].

(5) Connect the zeta polynomials of graphs with other polynomials associated with graphs and knots (e.g., Tutte, Alexander, and Jones polynomials). Papers exist, but the connection is mysterious to me. See Lin and Wang [75].

(6) Find more graph theoretic analogs of number theoretic results. The Galois theory of graph coverings allowed us to view Ihara zetas of graph covers as analogs of Dedekind zetas of extensions of number fields. We found analogs of the prime number theorem, the Frobenius automorphism, the Chebotarev density theorem, the explicit formulas of Weil from analytic number theory, and Siegel zeros. The analog of the ideal class group is the Jacobian of a graph and it has order equal to the number of spanning trees. See R. Bacher, P. de la Harpe, and Tatiana Nagnibeda [5] as well as Baker and Norine [7]. Are there graph analogs of regulators, Stark conjectures (see formula (10.6), Figure 16.2, and [118]), and class field theory for abelian graph coverings? Or, more simply, find a quadratic reciprocity law and fundamental units? It may help to remember that Ihara zeta functions are closer to zeta functions of function fields than to the zetas of number fields. See Rosen [104].

(7) Look at Avi Wigderson's website (http://www.math.ias.edu/~avi/) and find out what zig-zag products of graphs are. Does the definition depend on the labeling? Compute the zeta functions for these products. Are there any divisibility properties? Infinite families of regular expanders of arbitrary constant degree were obtained via the modified zig-zag product by Cristina Ballantine and Matthew Horton [9].

(8) Investigate the explicit formula for the Ihara zeta function. See [59] for applications of the regular graph explicit formula known as the Selberg trace formula. Find analogs of the applications of explicit formulas in number theory. See Lang [73] and Murty [91].

(9) Investigate the conjecture in Hoory, Linial, and Wigderson [55] that every d-regular graph X has a 2-covering Y such that, if A_Y is the adjacency matrix of Y,

$$\text{Spectrum } A_Y - \text{Spectrum } A_X \subset \left[-2\sqrt{d-1},\ 2\sqrt{d-1}\,\right].$$

See Proposition 18.17.

(10) Which basic invariants of the graph X can be determined by the Ihara zeta function? For example, $2|E|$ is the degree of the reciprocal of zeta. See Yaim Cooper [24], Debra Czarneski [33], Matthew Horton [57], [58], and Christopher Storm [125].

(11) Investigate the Angel, Friedman, and Hoory analog of the Alon conjecture for irregular graphs, given in [2].

References

[1] N. Alon and F. R. K. Chung, Explicit construction of linear sized tolerant networks, *Discrete Math.*, 72 (1989), 15–19.

[2] O. Angel, J. Friedman, and S. Hoory, The non-backtracking spectrum of the universal cover of a graph, preprint.

[3] T. Apostol, *Introduction to Analytic Number Theory*, Springer-Verlag, New York, 1984.

[4] M. Artin and B. Mazur, On periodic points, *Ann. Math.*, 81(2) (1965), 82–99.

[5] R. Bacher, P. de la Harpe, and T. Nagnibeda, The lattice of integral flows and the lattice of integral cuts on a finite graph, *Bull. Math. Soc. France*, 125 (1997), 167–198.

[6] Z. D. Bai, Circular law, *Ann. Prob.* 25 (1997), 494–529.

[7] M. Baker and S. Norine, Riemann–Roch and Abel–Jacobi theory on a finite graph, *Adv. Math.*, 215 (2007), 766–788.

[8] M. Baker and S. Norine, Harmonic morphisms and hyperelliptic graphs, *Int. Math. Res. Not.*, 15 (2009), 2914–2955.

[9] C. M. Ballantine and M. D. Horton, Infinite families of regular expanders of arbitrary constant degree obtained via the modified zig-zag product, preprint.

[10] R. Band, T. Shapira, and U. Smilansky, Nodal domains on isospectral quantum graphs: the resolution of isospectrality? *J. Phys. A: Math. Gen.*, 39 (2006), 13 999–14 014.

[11] L. Bartholdi, Counting paths in graphs, *Enseign. Math.*, 45 (1999), 83–131.

[12] H. Bass, The Ihara–Selberg zeta function of a tree lattice, *Intl. J. Math.*, 3 (1992), 717–797.

[13] T. Bedford, M. Keane, and C. Series (eds.), *Ergodic Theory, Symbolic Dynamics, and Hyperbolic Spaces*, Oxford University Press, Oxford, 1991.

[14] G. Berkolaiko, R. Carlson, S. Fulling, and P. Kuchment (eds.), *Proc. Joint Summer Research Conf. on Quantum Graphs and Their Applications, Contemporary Math.*, vol. 415, Amer. Math. Soc., Providence, 2006.

[15] N. Biggs, *Algebraic Graph Theory*, Cambridge University Press, London, 1974.

[16] O. Bohigas and M.-J. Giannoni, Chaotic motion and random matrix theories, In *Lecture Notes in Physics*, vol. 209, Springer-Verlag, Berlin, 1984, pp. 1–99.

[17] O. Bohigas, R. U. Haq, and A. Pandey, Fluctuation properties of nuclear energy levels and widths: comparison of theory with experiment, in K. H. Böckhoff (ed.), *Nuclear Data for Science and Technology*, Reidel, Dordrecht, 1983, pp. 809–813.

[18] B. Bollobás, *Graph Theory: An Introductory Course*, Springer-Verlag, New York, 1979.

[19] B. Bollobás, *Modern Graph Theory*, Springer-Verlag, New York, 1998.

[20] A. Borel and G. D. Mostow, Algebraic groups and discontinuous subgroups, in *Proc. Symp. Pure Math.*, IX, Amer. Math. Soc. Providence, 1966.

[21] M. G. Bulmer, *Principles of Statistics*, Dover, New York, 1979.

[22] U. Bunke and M. Olbrich, *Selberg Zeta and Theta Functions*, Akademie Verlag, Berlin, 1995.

[23] P. Buser, *Geometry and Spectra of Compact Riemann Surfaces*, Birkhäuser, Boston, 1992.

[24] Y. Cooper, Properties determined by the Ihara zeta function of a graph, *Electronic J. Combinatorics*, 16 (2009), #R84.

[25] F. R. K. Chung, Diameters and eigenvalues, *J. Amer. Math. Soc.*, 2 (1989), 187–196.

[26] F. R. K. Chung, *Spectral Graph Theory*, *CBMS Regional Conf. Series in Math.*, vol. 92, Amer. Math. Soc., Providence, 1996.

[27] F. R. K. Chung, Four proofs of Cheeger's inequality and graph partition algorithms, in *Proc. ICCM.*, 2007.

[28] F. R. K. Chung, L. Lu, and V. Vu, Spectra of random graphs with given expected degrees, *Proc. Natl Acad. Sci.*, 100(11) (2003), 6313–6318.

[29] F. R. K. Chung and S.-T. Yau, Coverings, heat kernels, and spanning trees, *Electron. J. Combinatorics*, 6 (1999), Research paper 12.

[30] B. Cipra, *What's Happening in the Mathematical Sciences, 1998–1999*, Amer. Math. Soc., Providence, 1999.

[31] B. Clair and S. Mokhtari-Sharghi, Zeta functions of discrete groups acting on trees, *J. Algebra*, 237 (2001), 591–620.

[32] D. M. Cvetković, M. Doob, and H. Sachs, *Spectra of Graphs: Theory and Application*, Academic Press, New York, 1979.

[33] D. L. Czarneski, Zeta functions of finite graphs, Ph.D. thesis, Louisiana State University, 2005.

[34] H. Davenport, *Multiplicative Number Theory*, Springer-Verlag, New York, 1980.

[35] G. Davidoff, P. Sarnak, and A. Valette, *Elementary Number Theory, Group Theory and Ramanujan Graphs*, Cambridge University Press, Cambridge, 2003.

[36] P. Diaconis, *Group Representations in Probability and Statistics*, Inst. Math. Statistics, Hayward CA, 1988.

[37] H. M. Edwards, *Riemann's Zeta Function*, Academic Press, New York, 1974.

[38] J. Elstrodt, Die Selbergsche Spurformel für kompakte Riemannsche Flächen, *Jber. d. Dt. Math.-Verein.*, 83 (1981), 45–77.

[39] J. Elstrodt, F. Grunewald, and J. Mennicke, *Groups acting on Hyperbolic Space*, Springer-Verlag, Berlin, 1998.

[40] D. W. Farmer, Modeling families of L-functions, arXiv:math.NT/0511107 v1, 4 November 2005.

[41] P. J. Forrester and A. Odlyzko, A nonlinear equation and its application to nearest neighbor spacings for zeros of the zeta function and eigenvalues of random matrices, in *Proc. Organic Math. Workshop*, invited articles, 1997, located at http://www.cecm.sfu.ca/~pborwein/

[42] J. Friedman, A proof of Alon's second eigenvalue problem, in *Mem. Amer. Math. Soc.*, vol. 195, no. 910, Amer. Math. Soc., Providence, 2008.

[43] J. Ginibre, *J. Math. Phys.*, 6 (1965), 440.

[44] V. L. Girko, Circular law, *Theory Prob. Appl.* 29 (1984), 694–706.

[45] C. Gordon, D. Webb, and S. Wolpert, Isospectral plane domains and surfaces via Riemannian orbifolds, *Inv. Math.*, 110 (1992), 1–22.

[46] R. I. Grigorchuk and A. Zuk, The Ihara zeta function of infinite graphs, the KNS spectral measure and integrable maps, in *Random Walks and Geometry*, Walter de Gruyter, Berlin, 2004, pp. 141–180.

[47] J. L. Gross and T. W. Tucker, *Topological Graph Theory*, Dover, Mineola, New York, 2001.

[48] D. Guido, T. Isola, and M. Lapidus, Ihara zeta functions for periodic simple graphs, arXiv: math/0605753v3 [math.OA] 5 March 2007.

[49] F. Haake, *Quantum Signatures of Chaos*, Springer-Verlag, Berlin, 1992.

[50] K. Hashimoto, Zeta functions of finite graphs and representations of p-adic groups, *Adv. Studies in Pure Math.*, vol. 15, Academic Press, New York, 1989, pp. 211–280.
[51] K. Hashimoto, Artin-type L-functions and the density theorem for prime cycles on finite graphs, *Intl J. Math.*, 3 (1992), 809–826.
[52] D. A. Hejhal, The Selberg trace formula and the Riemann zeta function, *Duke Math. J.*, 43 (1976), 441–482.
[53] D. A. Hejhal, J. Friedman, M. Gutzwiller, and A. Odlyzko (eds.), *Emerging Applications of Number Theory*, Institute for Mathematics and Its Applications, vol. 109, Springer-Verlag, New York, 1999.
[54] S. Hoory, A lower bound on the spectral radius of the universal cover of a graph, *J. Combin. Theory B*, 93 (2005), 33–43.
[55] S. Hoory, N. Linial, and A. Wigderson, Expander graphs and their applications, *Bull. Amer. Math. Soc.*, 43 (2006), 439–561.
[56] R. A. Horn and C. R. Johnson, *Topics in Matrix Analysis*, Cambridge University Press, Cambridge, 1991.
[57] M. D. Horton, Ihara zeta functions of irregular Graphs, Ph.D. Thesis, University of California, San Diego, 2006.
[58] M. D. Horton, Ihara zeta functions of digraphs, *Linear Alg. Appl.*, 425 (2007), 130–142.
[59] M. D. Horton, D. Newland, and A. A. Terras, The contest between the kernels in the Selberg trace formula for the $(q+1)$-regular tree, in J. Jorgenson and L. Walling (eds.), *The Ubiquitous Heat Kernel, Contemp. Math.*, vol. 398, Amer. Math. Soc., Providence, 2006, pp. 265–293.
[60] M. D. Horton, H. M. Stark, and A. A. Terras, What are zeta functions of graphs and what are they good for?, in G. Berkolaiko, R. Carlson, S. A. Fulling, and P. Kuchment (eds.), *Quantum Graphs and Their Applications*, Contemp. Math., vol. 415 (2006), pp. 173–190.
[61] M. D. Horton, H. M. Stark, and A. A. Terras, Zeta functions of weighted graphs and covering graphs, in *Analysis on Graphs*, *Proc. Symp. Pure Math.*, vol. 77, Amer. Math. Soc., Providence, 2008.
[62] Y. Ihara, On discrete subgroups of the two by two projective linear group over p-adic fields, *J. Math. Soc. Japan*, 18 (1966), 219–235.
[63] K. Ireland and M. Rosen, *A Classical Introduction to Modern Number Theory*, Springer-Verlag, New York, 1982.
[64] H. Iwaniec and E. Kowalski, *Analytic Number Theory*, Amer. Math. Soc., Providence, 2004.
[65] M. Kac, Can you hear the shape of a drum?, *Amer. Math. Monthly*, 73 (1966), 1–23.
[66] M.-H. Kang, W.-C. W. Li, and C.-J. Wang, The zeta functions of complexes from PGL(3): a representation-theoretic approach, arXiv:0809.1401v2 [math.NT], 19 November 2008.
[67] S. Katok, *Fuchsian Groups*, University of Chicago Press, Chicago, 1992.
[68] N. Katz and P. Sarnak, *Random Matrices, Frobenius Eigenvalues and Monodromy*, Amer. Math. Soc., Providence, 1999.
[69] N. Katz and P. Sarnak, Zeroes of zeta functions and symmetry, *Bull. Amer. Math. Soc.*, 36(1) (1999), 1–26.
[70] R. Koetter, W.-C. W. Li, P. O. Vontobel, and J. L. Walker, Pseudo-codewords of cycle codes via zeta functions, arXiv:cs/0502033v1, 6 February 2005.
[71] R. Koetter, W.-C. W. Li, P. O. Vontobel, and J. L. Walker, Characterizations of pseudo codewords of LDPC codes, *Adv. Math.*, 217 (2007), 205–229.
[72] M. Kotani and T. Sunada, Zeta functions of finite graphs, *J. Math. Sci. Univ. Tokyo*, 7 (2000), 7–25.
[73] S. Lang, *Algebraic Number Theory*, Addison-Wesley, Reading MA., 1968.
[74] P. LeBoeuf, Random matrices, random polynomials, and Coulomb systems, arXiv: cond-mat/9911222v1 [cond-mat.stat-mech], 15 November 1999.

[75] X.-S. Lin and Z. Wang, Random walk on knot diagrams, colored Jones polynomial and Ihara–Selberg zeta function, in J. Gilman *et al.* (eds.), *Knots, Braids, and Mapping Class Groups – Papers Dedicated to Joan S. Birman, Proc. Conf. in Low Dimensional Topology in Honor of Joan S. Birman's 70th Birthday, IP Stud. Adv. Math.*, vol. 24, Amer. Math. Soc., Providence, 2006, pp. 107–121.

[76] D. Lorenzini, *An Invitation to Arithmetic Geometry, Grad. Studies in Math.*, vol. 9, Amer. Math. Soc., Providence, 1997.

[77] A. Lubotzky, *Discrete Groups, Expanding Graphs and Invariant Measures*, Birkhäuser, Basel, 1994.

[78] A. Lubotzky, Cayley graphs: eigenvalues, expanders and random walks, in *Surveys in Combinatorics, London Math. Soc. Lecture Notes*, vol. 218, Cambridge University Press, Cambridge, 1995, pp. 155–189.

[79] A. Lubotzky, R. Phillips, and P. Sarnak, Ramanujan graphs, *Combinatorica*, 8 (1988), 261–277.

[80] A. Lubotzky, B. Samuels, and U. Vishne, Isospectral Cayley graphs of some finite simple groups, *Duke Math. J.*, 135 (2006), 381–393.

[81] B. Malmskog and M. Manes, Almost divisibility in the Ihara zeta functions of certain ramified covers of $q+1$-regular graphs, preprint.

[82] G. A. Margulis, Explicit group-theoretic constructions of combinatorial schemes and their applications in the construction of expanders and concentrators, *Prob. Inf. Transm.*, 24 (1988), 39–46.

[83] W. S. Massey, *Algebraic Topology: An Introduction*, Springer-Verlag, New York, 1967.

[84] B. D. McKay, The expected eigenvalue distribution of a large regular graph, *Linear Alg. Appl.*, 40 (1981), 203–216.

[85] M. L. Mehta, *Random Matrices and the Statistical Theory of Energy Levels*, Academic Press, New York, 1990.

[86] S. J. Miller and R. Takloo-Bighash, *An Invitation to Modern Number Theory*, Princeton University Press, Princeton, 2006.

[87] S. J. Miller, T. Novikoff and A. Sabelli, The distribution of the largest non-trivial eigenvalues in families of random regular graphs, *Experimental Math.* 17(2) (2008), 231–244.

[88] H. Mink, *Nonnegative Matrices*, Wiley, New York, 1988.

[89] H. Mizuno and I. Sato, The scattering matrix of a graph, *Electronic J. Combinatorics*, 15 (2008), #R96, 1–16.

[90] H. L. Montgomery, The pair correlation of zeros of the zeta function, in *Proc. Symp. Pure Math.*, vol. 24, Amer. Math. Soc., Providence, 1973, pp. 181–193.

[91] R. Murty, *Problems in Analytic Number Theory*, Springer-Verlag, New York, 2001.

[92] H. Nagoshi, On arithmetic infinite graphs, *Proc. Japan Acad. A*, 76 (2000), 22–25.

[93] H. Nagoshi, Spectra of arithmetic infinite graphs and their application, *Interdiscip. Inform. Sci.*, 7(1) (2001), 67–76.

[94] H. Nagoshi, Spectral theory of certain arithmetic graphs, in *Contemp. Math.*, vol. 347, Amer. Math. Soc., Providence, 2004, pp. 203–220.

[95] D. Newland, Kernels in the Selberg Trace Formula on the k-regular tree and zeros of the Ihara zeta function, Ph.D. thesis, University of California, San Diego, 2005.

[96] T. J. Osborne and S. Severini, Quantum algorithms and covering spaces, arXiv: quant-ph/0403127, v. 3, 11 May 2004.

[97] S. J. Patterson, *An Introduction to the Theory of the Riemann Zeta Function*, Cambridge University Press, Cambridge, 1988.

[98] S. Patterson, Review of *Selberg Zeta and Theta Functions* by Bunke and Olbrich, *Bull. Amer. Math. Soc.*, 34 (1997), 183–186.

[99] R. Perlis, On the equation $\zeta_K(s) = \zeta_{K'}(s)$, *J. Number Theory*, 9 (1977), 342–360.

[100] T. Petrillo, Ph.D. thesis, University of California, San Diego, 2010.

[101] V. Pless, *Introduction to the Theory of Error-Correcting Codes*, Wiley, New York, 1989.

[102] A. Odlyzko, On the distribution of spacings between zeros of the zeta function, *Math. Comput.*, 48 (1987), 273–308.

[103] R. Reeds, Zeta functions on Kronecker products of graphs, http://www.rose-hulman.edu/mathjournal/archives/2006/vol7-n1/paper3/v7n1-3pd.pdf.

[104] M. Rosen, *Number Theory in Function Fields*, Springer-Verlag, New York, 2002.

[105] M. Rubinstein, Evidence for a spectral interpretation of zeros of L-functions, Ph.D. thesis, Princeton University, 1998.

[106] Z. Rudnick, some problems in "quantum chaos" or statistics of spectra, Park City Lectures, 2002.

[107] Z. Rudnick, What is quantum chaos? *Not. Amer. Math. Soc.*, 55(1) (2008), 32–34.

[108] D. Ruelle, *Dynamical Zeta Functions for Piecewise Monotone Maps of the Interval,* Amer. Math. Soc., Providence, 1994.

[109] P. Sarnak, *Some Applications of Modular Forms*, Cambridge University Press, Cambridge, 1990.

[110] P. Sarnak, Arithmetic quantum chaos, in *Proc. Israel Math. Conf.*, vol. 8, Amer. Math. Soc., 1995, pp. 183–236.

[111] A. Selberg, Harmonic analysis and discontinuous groups in weakly symmetric Riemannian spaces with applications to Dirichlet series, *J. Ind. Math. Soc.*, 20 (1954), 47–87.

[112] A. Selberg, *Collected Papers*, vol. I, Springer-Verlag, New York, 1989.

[113] J.-P. Serre, *Trees*, Springer-Verlag, New York, 1980.

[114] S. Skiena, *Implementing Discrete Mathematics: Combinatorics and Graph Theory with Mathematica*, Addison-Wesley, Redwood City CA, 1990.

[115] U. Smilansky, Quantum chaos on discrete graphs, *J. Phys. A: Math. Theor.*, 40 (2007), F621–F630.

[116] H. M. Stark, Some effective cases of the Brauer–Siegel theorem, *Inventiones Math.*, 23 (1974), 135–152.

[117] H. M. Stark, Hilbert's twelfth problem and L series, *Bull. Amer. Math. Soc.*, 83 (1977), 1072–1074.

[118] H. M. Stark, Galois theory, algebraic number theory and zeta functions, in M. Waldschmidt *et al.* (eds.), *From Number Theory to Physics*, Springer-Verlag, Berlin, 1992, pp. 313–393.

[119] H. M. Stark and A. A. Terras, Zeta functions of finite graphs and coverings, *Adv. Math.*, 121 (1996), 124–165.

[120] H. M. Stark and A. A. Terras, Zeta functions of finite graphs and coverings, II, *Adv. Math.*, 154 (2000), 132–195.

[121] H. M. Stark and A. A. Terras, Zeta functions of graph coverings, in *DIMACS Series in Discrete Math. and Theor. Comp. Sci.*, vol. 64, 2004, pp. 199–212.

[122] H. M. Stark and A. A. Terras, Zeta functions of finite graphs and coverings, III, *Adv. Math.*, 208 (2007), 467–489.

[123] P. Stevenhagen and H. W. Lenstra, Chebotarëv and his density theorem, *The Math. Intelligencer*, 18(2) (1996), 26–37.

[124] C. K. Storm, The zeta function of a hypergraph, *Electronic J. Combinatorics* 13 (2006), #R84.

[125] C. K. Storm, Some graph properties determined by edge zeta functions, arXiv:0708.1923v1 [math.CO] 14 August 2007.

[126] A. Strömbergsson, On the zeros of L-functions associated to Maass waveforms, *Intl Math. Res. Notices*, 15 (1999), 839–851.

[127] T. Sunada, Riemannian coverings and isospectral manifolds, *Ann. Math.*, 121 (1985), 169–186.

[128] T. Sunada, L-functions in geometry and some applications, in *Lecture Notes in Math.*, vol. 1201, Springer-Verlag, New York, 1986, pp. 266–284.
[129] T. Sunada, Fundamental groups and Laplacians, in *Lecture Notes in Math.*, vol. 1339, Springer-Verlag, New York, 1988, pp. 248–277.
[130] T. Sunada, in *The Discrete and the Continuous, Proc. Sugaku Seminar* (in Japanese), vol. 40, Nippon Hyouronsha, Tokyo, 2001, pp. 48–51.
[131] T. Tao and V. Vu, From the Littlewood–Offord problem to the circular law: universality of the spectral distributions of random matrices, *Bull. Amer. Math. Soc.*, 46 (2009), 377–396.
[132] A. Terras, *Harmonic Analysis on Symmetric Spaces and Applications, vols. I,II*, Springer-Verlag, New York, 1985, 1988.
[133] A. Terras, *Fourier Analysis on Finite Groups and Applications*, Cambridge University Press, Cambridge, 1999.
[134] A. Terras, Finite quantum chaos, *Amer. Math. Monthly*, 109 (2002), 121–139.
[135] A. Terras, Finite models for arithmetical quantum chaos, in *IAS/Park City Math. Series*, vol. 12, 2007, pp. 333–375.
[136] A. Terras and D. Wallace, Selberg's trace formula on k-regular trees and applications, *Intl J. Math. Math. Sci.*, 8 (2003), 501–526.
[137] C. Trimble, Ph.D. thesis, Some special functions on p-adic and finite analogues of the Poincaré upper half plane, University of California, San Diego, 1993.
[138] A. B. Venkov and A. M. Nikitin, The Selberg trace formula, Ramanujan graphs and some problems of mathematical physics, *Petersburg Math. J.*, 5 (1994), 419–484.
[139] Vignéras, M.-F., L'équation fonctionelle de la fonction zêta de Selberg de la groupe modulaire PSL$(2, \mathbb{Z})$, *Astérisque*, 61 (1979), 235–249.
[140] N. Ya. Vilenkin, *Combinatorics*, Academic Press, New York, 1971.
[141] A. Weil, Sur les "formules explicites" de la théorie des nombres premiers, in *Comm. Sém. Math. Univ. Lund 1952, Tome supplémentaire*, 1952, pp. 252–265.
[142] H. Weyl, Über die Gleichverteilung von Zahlen mod. Eins, *Math. Ann.*, 77 (1916), 313–352.
[143] E. P. Wigner, Random matrices in physics, *SIAM Rev.*, 9(1) (1967), 1–23.

Index

adjacency matrix, 10
algebraic variety, 27
Alon conjecture, 17
analytic density, 196
Artin–Ihara L-function properties, 154
Artinized
  adjacency matrix, 156
  edge adjacency matrix, 149, 165
  path matrix, 179
  start, terminal, and J matrices, 170

backtrack, 11
Bass proof of Artin–Ihara three-term determinant formula, 89, 173
boundary of a set of vertices, 71
bounds
  on poles of Ihara zeta, 94–95
bouquet of loops, 14

Catalan number, 69–70
centralizer of element of group, 15
character
  of cyclic group, 63–64
  of representation, 146
character table for S_3, 148
Chebotarev density theorem, 194–199
closed path, 11
complexity of a graph, 78
conjectured inequality between ρ_X, R_X, and $\overline{d_X}$, 54
conjugacy class in group, 14
conjugate intermediate graph, 122
  equivalent to covering isomorphic graph, 122
conjugate prime, 131
conjugate subgroup, 122
constructing graph covers using the Frobenius automorphism, 142, 190, 191

covering isomorphic graphs, 118
covering isomorphism, 118
covering map π, 20, 106
covering of graphs, 20, 106
cycle, 11
cycle code, 213
cycle graph, 10

decomposition group, 137
Dedekind conjecture, 115
degree
  of representation, 144
  of vertex, 10
δ, 201
Δ_X, 21
determinant formula, three term,
  for Artin–Ihara zeta, 156
  for Ihara zeta, 17
determinant formula, two term,
  cohomological, for zeta of algebraic variety, 27
  for edge zeta, 84
  for Ihara zeta, 28, 86
  for path Artin L-function, 179
  for path zeta, 100
  for Artin–Ihara L-function, 149
  with edge adjacency matrix W_1, 28
dichotomy of quantum chaos, 39
direct sum of representations, 145
Dirichlet theorem on primes in progressions, 7
divisibility of zetas of covers, 110

edge adjacency matrix W_1, 28
  irreducible, 93
edge Artin L-function properties, 165
edge Artin matrix, 149, 165
edge matrix W, 83
edge norm, 83

elementary reduction operations on
fundamental group, 99
equal intermediate graphs, 118
equidistributed sequence, 39
equivalence class of closed paths, 12
equivalent representations, 145
error-correcting code, 211
Euclid's fifth postulate, 23
Euler characteristic, 16
Euler product, 3
expander family, 71
expansion ratio, 71
experiment
Mathematica on random graphs, 57
Matlab
on abelian cover, 225
on random cover, 224
on random graph, 88, 221
Newland using Mathematica on random
regular graphs, 40–41
ramification, 115, 116
explicit formulas, 4, 216–217
extended Riemann hypothesis (ERH), 7

factorization
of edge zeta of a normal cover, 165
of Ihara zeta of a normal cover, 155
of path zeta of a normal cover, 184
facts about edge adjacency matrix, 91
$f(D, Y/X)$, residual degree of D, 128
fission, 86
Fourier inversion, 144
Fourier transform, 144
fractional linear transformation, 22
free group, 13
freely homotopic cycles, 15
Frobenius
automorphism of graph, properties, 137
automorphism of a path with respect to a
Galois cover, 135
character formula, 146
morphism of algebraic variety, 27
normalized automorphism of a path with
respect to a Galois cover, 133
reciprocity law, 147
functional equation
regular graph, 50
Riemann zeta, 3
fundamental domain of modular group, 23
fundamental group, 13, 99
fundamental theorem of graph
Galois theory, 118

$\widehat{G}$, set of all irreducible unitary representations
of group G, 145
Galois group action on sheets of cover, 110
Galois group $G(Y/X)$, 109
Gauss sum, 64
Gaussian orthogonal ensemble (GOE), 34
Gaussian unitary ensemble (GUE), 4, 35
$g(D, Y/X)$, number of primes above C in X if
$D|C$, 128
generalized Riemann hypothesis
(GRH), 54
geodesic in upper half plane, 22, 24
graph
bipartite, 47
bouquet of loops, 14
Cayley, 50
complete, 10
cycle, 10
dumbbell, 86
Euclidean, 65
Harold and Audrey, 192
K_4, tetrahedron, 15
$K_4 - e$, 15
Lubotzky, Phillips, and Sarnak, 67
normal for a cycle code, 213
Paley, 64
regular, 10
simple, 10
Tanner, 212
graph automorphism, 109
graph prime counting function, 21
group
free, 13
fundamental, 13
group representation, 144

hypotheses, the usual, 10

induced representation, 145
induction property
of Artin–Ihara L-functions, 154
of edge Artin L-functions, 166, 175
of path Artin L-functions, 183
inequality for ρ, p, q, R, 54, 96
inflated graph, 45, 206
intermediate covering, 117
irreducible matrix, 90
irreducible representation, 144
isospectral non-isomorphic graphs, 188
Buser's example, 187
Stark and Terras example, 192

K_4, 16
Kloosterman sum, 66

L-function
Artin–Ihara, 148
Artin of number fields, 134
Dirichlet, 6, 7
edge Artin, 164
path Artin, 178

labeling edges
left out of spanning tree, 98
of a graph, 11
Landau's theorem, 13
Laplace operator for Poincaré arc length, 22
lemma
expander mixing, 71
$R_Y = R_X$, 195
length $\nu(C)$ of path C, 12
level spacings picture, 31, 33, 35, 41, 220, 225, 226
lift, 109

Möbius function, 76
Möbius inversion formula, 76
Markov transition matrix, 61
matrix identities with S and T, 88
with representations, 170
modular group, 23

naive Ramanujan inequality, 53
neighborhood of a vertex, 106
Newland's experiments, 40, 41
N_m, number of closed paths of length m without backtracking and tails, 29
node, 200
non-Euclidean triangle, 23
normal or Galois covering, 109
number of primes D above C, 128

orthogonality relations, 147

parity check matrix, 211
path matrix Z, 99
path norm, 100
permutation matrix, 90
$\pi(n)$, prime counting function, 21
Plancherel measure, 39
Poincaré arc length, 22
Poincaré upper half plane, 22
prime ideal theorem, 7
prime in a graph, 12
prime number theorem
graph theory version, 21, 75
ordinary integer version, 4
prime or primitive path in graph, 12
primitive conjugacy class, 15
probability vector, 61
properties
of Artin–Ihara L-functions, 154
of edge Artin L-functions, 165
of path Artin L-functions, 179

quantum graph, 192

R, radius of circle of convergence of Ihara zeta, 13
Ramanujan
graph, 17
irregular graph, 53
ramification, 8, 24, 115
rank of free group, 13
Rayleigh quotient, 94
reduced word in fundamental group, 99
k-regular graph, 10
relation between δ and Δ, 201
representation
direct sum, 145
equivalent, 145
induced, 145
of cyclic group, 144
right regular, 145
trivial, 145
unitary, 144
residual degree $f(D, Y/X)$, 128
ρ_X, 52
ρ'_X, 52
Riemann hypothesis (RH)
graph theory, 53
regular graph, 48
Riemann zeta, 3
weak graph theory, 54

S_3 cover of tetrahedron minus an edge, 124
Seifert–Van Kampen theorem, 13
shape of a drum, 186, 192
sheets of a covering, 107
Siegel poles, 201
Siegel zero
Dedekind zeta, 5
spanning tree, 13, 107
specialization, 84
edge zeta
to Hashimoto edge zeta, 84
to Ihara zeta, 84
to weighted zeta, 84
for induction property of path Artin L-functions, 181
of path zeta to edge zeta, 100
spectrum adjacency matrix, regular graph properties, 47
splitting of primes
in coverings, 128
in cube cover K_4, 130, 131
in non-normal cubic cover of the tetrahedron minus an edge, 132
in non-normal cubic extension of $\mathbb{Q}$, 129
in quadratic extension of $\mathbb{Q}$, 8
start matrix, 87
subshift of finite type, 29
symmetric generating set S for Cayley graph $X(G, S)$, 50

tail, 11
tensor product, 170
terminal matrix, 87
tessellation of upper half plane, 24
tetrahedron, 16
theorem
  Alon–Boppana, 68
  Bowen–Lanford, 30
  constructing graph covers using the
    Frobenius automorphism, 142
  Fan Chung, on graph diameters, 73
  fundamental theorem of graph
    Galois theory, 118
  graph theory prime number theorem, 75
  Hoory inequality, 53
  Ihara generalized by Bass, Hashimoto,
    etc., 17
  Koetter, Li, Vontobel, and Walker, 214
  Kotani and Sunada, 52
    proof of, 95
  McKay, 40
  Perron–Frobenius, 94
  random walker gets lost, 62
  right regular representation is mother of all
    representations, 145
transitivity property of residual degrees,
    131–132
tree, 13
trivial representation, 145
types of fractional linear transformation, 24

unique lifts to covers, 108
universal covering tree, 15
unramified covering graph, 20

weighted length of a path in
    a graph, 45
Weil conjectures, 27
Weil's proof of the Riemann hypothesis
    for zetas of curves, 66, 67
Wigner semicircle distribution, 33
Wigner surmise, 34, 218, 220

$X(G, S)$, Cayley graph, 50
X-specialized edge matrix, 165

zeta function
  algebraic variety, 27
  Artin–Mazur, 28
  Bartholdi, 84
  Dedekind, 5–6
  edge, 83
  Epstein, 9
  function field, 5
  Hashimoto edge, 84
  Ihara, 12, 16
  path, 100
  Riemann, 3
  Ruelle, 28
  Selberg, 25
  weighted graph, 45

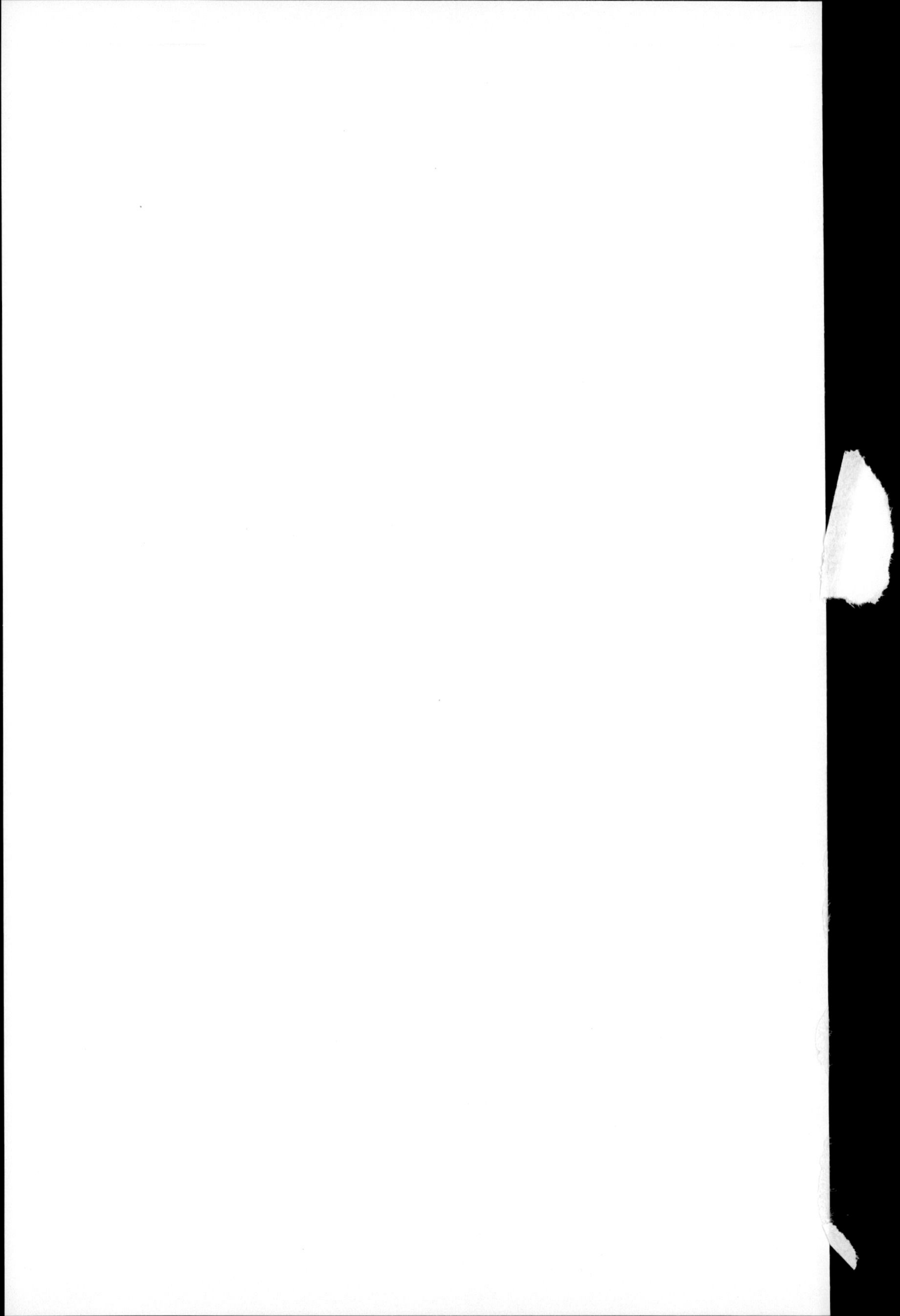

Printed in the United States
By Bookmasters